# UNSPEAKABLE HORROR

# UNSPEAKABLE HORROR

## THE DEADLIEST SHARK ATTACKS IN MARITIME HISTORY

WRITTEN AND EDITED
BY JOSEPH B. HEALY

SKYHORSE PUBLISHING

Skyhorse Publishing books may be purchased in bulk at special discounts for sales promotion, corporate gifts, fund-raising, or educational purposes. Special editions can also be created to specifications. For details, contact the Special Sales Department, Skyhorse Publishing, 307 West 36th Street, 11th Floor, New York, NY 10018 or info@skyhorsepublishing.com.

Skyhorse® and Skyhorse Publishing® are registered trademarks of Skyhorse Publishing, Inc.®, a Delaware corporation.

Visit our website at www.skyhorsepublishing.com.

10 9 8 7 6 5 4 3 2 1

Library of Congress Cataloging-in-Publication Data is available on file.

Cover design by Tom Lau
Cover photo credit: iStockphoto

Print ISBN: 978-1-5107-1935-4
Ebook ISBN: 978-1-5107-1936-1

Printed in the United States of America.

# CONTENTS

"The shark is nature's perfect design for survival in the sea."

—Robert Reid, in *Shark!: Killer Tales from the Dangerous Depths*

"It wasn't the jaws that mesmerized me, it was those protrusions on either side of its head. They belonged to a creature to whom I was nothing but an easy kill."

—*Stephen H. Foreman, describing a hammerhead shark in "Blackfish and Hammerheads of Barrouallie"*

# INTRODUCTION:
# CONTEMPLATING MASS
# DEATH AT SEA

Thoughts on *Unspeakable Horror: The Worst Shark
Attacks in Maritime History*
                    *—By Joseph B. Healy*

The story of the USS *Indianapolis* from World War II is well known and gets a fair representation in this book. Captain Quint delivered a suspenseful soliloquy in the below-deck galley scene of *Jaws*, before the movie's climactic great-white attack that crippled the *Orca* and sent Quint sliding into the shark's jaws, and he disappeared spitting blood. The story of the *Indianapolis*: After delivering crucial components of the atomic bomb that would destroy the Japanese city Hiroshima in 1945, the *Indianapolis* was sunk by a Japanese submarine near the Philippine Sea. Of the nearly 1,200 men aboard, 900 survived the torpedoing, spilling into the sea. Whitetip sharks began attacking the next morning, and after four days only 317 sailors remained alive for the rescue operation.

Less famous are the many stories of ships sinking in shark-infested waters with gruesome results. Such as the *Cape San Juan*, a US troop transport ship that was torpedoed by a Japanese submarine in the South Pacific Ocean near the Fiji Islands. Or the HMS *Birkenhead*, which sunk off Danger Point, South Africa, in 1852. In 1927, the luxury Italian cruise liner *Principessa Mafalda* sank ninety miles off the coast of Albrohos Island while heading to Porto Seguro, Brazil. More than 300 who initially survived the wreck were killed, many by sharks. In 1909, the French steamer *La Seyne* collided with British India Steamship Co. liner *Onda* near Singapore, twenty-six miles from land. One hundred and one people were eventually killed by sharks. These and more tragedies and disasters are detailed in the pages of this book.

In the water, human intelligence is no match for a shark's instincts to feed. Sharks are born to kill and eat. Eating and mating—surviving and procreating—is about all they do. They are all muscle and flexible cartilage, streamlined with nature's design for fleet swimming and fierce attacks, mouths of razor teeth, and heightened senses. They detect distress, smell blood—and attack. Marine disasters such as those mentioned above result in humans becoming prey, floating in inner space as sharks like haunting specters swim below, staging to attack, kill, and eat. Helpless to save yourself—floating and waiting, watching the malevolent creatures circle, knowing what will happen . . . a sudden swirl of water, a cloud of blood, the searing pain . . . until there is no more. This is unspeakable horror.

We shudder when we think about sharks, and just as with a scary ghost story, we revel in the fear. We don't think too much about sharks when we're at the beach; that would ruin our day. But we know they could be there, looking for a meal.

In classic literature, we find sharks, on some occasions. The novel *Jaws* by Peter Benchley is perhaps the most famous, though more so the movie version directed by Steven Spielberg. But let's not forget Santiago fighting the sharks in Hemingway's *The Old Man and the Sea*—and his lesson that man can be destroyed but not defeated. The master of outdoor adventure Jack London also brings sharks into his work, in one of his stories that's lesser known than his Klondike wolf stories called "The Heathen." London sailed quite a lot, lived on the sea, and knew his subject well. I'll let him tell the rest.

# FROM "THE HEATHEN"

### —By Jack London

Stranded after a hurricane-caused shipwreck on an island in the South Seas with a native man who saved him from drowning in the wreck, the narrator of Jack London's classic story "The Heathen" becomes his savior's friend for life. They "perform the ceremony of exchanging names," and we learn the narrator is Otoo and calls the native man Charley. The narrator says of the title character, "And he knew nothing of common Christian morality. All the people on Bora Bora were Christians; but he was a heathen, the only unbeliever on the island, a gross materialist, who believed that when he died he was dead. He believed merely in fair play and square dealing. Petty meanness, in his code, was almost as serious as wanton homicide; and I do believe that he respected a murderer more than a man given to small practices." For seventeen years, they stayed together in a brotherhood—until, as London writes: ". . . the end came, as the end must come to all human associations." The end for Otoo/Charley, the heathen, was a horrible shark attack during which he sacrificed himself to protect his now compatriot or brother. What follows is an excerpt from the story, first published in *Everyman's Magazine* in 1910.

*It occurred in the Solomons, where our wildest work had been done in the wild young days, and where we were once more—principally on a holiday, incidentally to look after our holdings on Florida Island and to look over the pearling possibilities of the Mboli Pass. We were lying at Savo, having run in to trade for curios.*

*Now, Savo is alive with sharks. The custom of the woolly-heads of burying their dead in the sea did not tend to discourage the sharks from making the adjacent waters a hangout. It was my luck to be coming aboard in a tiny, overloaded, native canoe, when the thing capsized. There were four woolly-heads and myself in it, or rather, hanging to it. The schooner was a hundred yards away.*

*I was just hailing for a boat when one of the woolly-heads began to scream. Holding on to the end of the canoe, both he and that portion of the canoe were dragged under several times. Then he loosed his clutch and disappeared. A shark had got him.*

*The three remaining n\*\*\*\*\*s tried to climb out of the water upon the bottom of the canoe. I yelled and cursed and struck at the nearest with my fist, but it was no use. They were in a blind funk. The canoe could barely have supported one of them. Under the three it upended and rolled sidewise, throwing them back into the water.*

*I abandoned the canoe and started to swim toward the schooner, expecting to be picked up by the boat before I got there. One of the n\*\*\*\*\*s elected to come with me, and we* *swam along silently, side by side, now and again putting our faces into the water and peering about for sharks. The screams of the man who stayed by the canoe informed us that he was taken. I was peering into the water when I saw a big shark pass directly beneath me.*

*He was fully sixteen feet in length. I saw the whole thing. He got the woolly-head by the middle, and away he went, the poor devil, head, shoulders, and arms out of the water all the time, screeching in a heart-rending way. He was carried along in this fashion for several hundred feet, when he was dragged beneath the surface.*

*I swam doggedly on, hoping that that was the last unattached shark. But there was another. Whether it was one that had attacked the natives earlier, or whether it was one that had made a good meal elsewhere, I do not know. At any rate, he was not in such haste as the others. I could not swim so rapidly now, for a large part of my effort was devoted to keeping track of him. I was watching him when he made his first attack. By good luck I got both hands on his nose, and, though his momentum nearly shoved me under, I managed to keep him off. He veered clear, and began circling about again. A second time I escaped him by the same maneuver. The third rush was a miss on both sides. He sheered at the moment my hands should have landed on his nose, but his sandpaper hide (I had on a sleeveless undershirt) scraped the skin off one arm from elbow to shoulder.*

*By this time I was played out, and gave up hope. The schooner was still two hundred feet away. My face was in the water, and I was watching him maneuver for another attempt, when I saw a brown body pass between us. It was Otoo.*

"Swim for the schooner, master!" he said. And he spoke gayly, as though the affair was a mere lark. "I know sharks. The shark is my brother."

I obeyed, swimming slowly on, while Otoo swam about me, keeping always between me and the shark, foiling his rushes and encouraging me.

"The davit tackle carried away, and they are rigging the falls," he explained, a minute or so later, and then went under to head off another attack.

By the time the schooner was thirty feet away I was about done for. I could scarcely move. They were heaving lines at us from on board, but they continually fell short. The shark, finding that it was receiving no hurt, had become bolder. Several times it nearly got me, but each time Otoo was there just the moment before it was too late. Of course, Otoo could have saved himself any time. But he stuck by me.

"Goodbye, Charley! I'm finished!" I just managed to gasp. I knew that the end had come, and that the next moment I should throw up my hands and go down.

But Otoo laughed in my face, saying: "I will show you a new trick. I will make that shark feel sick!"

He dropped in behind me, where the shark was preparing to come at me.

"A little more to the left!" he next called out. "There is a line there on the water. To the left, master—to the left!"

I changed my course and struck out blindly. I was by that time barely conscious. As my hand closed on the line I heard an exclamation from on board. I turned and looked. There was no sign of Otoo. The next instant he broke surface. Both hands were off at the wrist, the stumps spouting blood.

*"Otoo!" he called softly. And I could see in his gaze the love that thrilled in his voice.*

*Then, and then only, at the very last of all our years, he called me by that name.*

*"Goodbye, Otoo!" he called.*

*Then he was dragged under, and I was hauled aboard, where I fainted in the captain's arms.*

*And so passed Otoo, who saved me and made me a man, and who saved me in the end. We met in the maw of a hurricane, and parted in the maw of a shark, with seventeen intervening years of comradeship, the like of which I dare to assert has never befallen two men, the one brown and the other white. If Jehovah be from His high place watching every sparrow fall, not least in His kingdom shall be Otoo, the one heathen of Bora Bora.*

It's curious that Bora Bora is a setting in London's story, as so-called primitive cultures such as those in French Polynesia or the Solomon Islands in the South Pacific are known for deifying sharks. Sharks can inspire devotion among human cultures—though these beliefs are often grounded in violence and brutality. A story in the *Washington Post* from 1909 tell us more:

*In view of the wide distribution of sharks and their strength and ferocity, qualities which appealed to the savage mind, it is not strange that the cult of shark worship should have arisen. This worship is especially common in the South Seas, where sharks are very numerous.*

*In the Solomon Islands, living sacred objects are chiefly sharks, alligators, snakes. Sharks are in all these islands* very often thought to be the abode of ghosts, as natives will at times before their death announce that they will appear as sharks. Afterward any shark remarkable for size or color which is observed to haunt a certain shore or rock is taken to be some one's ghost, and the name of the deceased is given to it.

Such a one was Sautahimatawa at Ulawa, a dreaded man-eater, to which offerings of porpoise teeth were made. At Saa certain food, such as cocoanuts from certain trees, is reserved to feed such a ghost shark, and there are certain men of whom it is known that after death they will be sharks. These, therefore, are allowed to eat such food in the sacred place. In Saa and in Ulawa when a sacred shark had attempted to seize a man and he had escaped, the people would be so much afraid of the shark's anger that they would throw the man back in the sea to be drowned. These sharks also were thought to aid in catching bonito, for taking which supernatural power was necessary. In the Banks Islands a shark may be a tangaros, a sort of familiar spirit, or the abode of one. Some years ago Manurwar, son of Mala, the chief man in Vanua Lava, had such a shark. He had given money to a Manwo man to send it to him. It was very tame and would come up to him when he went down to the beach at Nawono and follow along in the surf as he walked along the shore. In the New Hebrides some men have the power, the natives believe, of changing themselves into sharks.

The Samoan native believed that his gods appeared in some visible incarnation, and the particular thing in which it was in the habit of appearing was to him an object of veneration. Many worshiped the shark in this way, and while they would freely partake of the gods of others, they *felt that death would be the penalty should they eat their own god. The god was supposed to avenge the insult by taking up his abode in the offender's body and causing to generate there the very thing which he had eaten until it produced death. . . . A shark named Moaalli was famous as the marine god of Molokai and Oahu. Many temples were built on promontories in his honor, and to them the first fruits of the fisherman's labors were dedicated. When victims were required to be sacrificed in honor of this dog, or he was supposed to be hungry, the priests would sally forth and ensnare with a rope any one they could catch. The victim was immediately strangled, cut in pieces, and thrown to the voracious animal.*

*Ukanipo was the shark god of Hawaii. He seems to have been of a compassionate nature at times, as there are extant several traditions showing kindnesses he had done to certain of his devotees, especially loves in distress.*

*All the shark gods were not beneficent, however. Apukohai and Uhumakaikai were evil shark gods who infested the waters of Kauai, and the fisherman were compelled to propitiate them with offerings.*

*Should a fisherman, by an unlucky accident, injure or destroy a shark held sacred by his family, he was bound to make a feast to the god.*

*Several of the African coast tribes worship the shark. Three or four times in the year they celebrate the festival of the shark, which is done in this wise: They all row out in their boats to the middle of the river, where they invoke, with the strangest ceremonies, the protection of the great shark. They offer to him poultry and goats in order to satisfy his sacred appetite. But this is nothing; an infant is every year sacrificed to the monster, which has been feted and nourished for the sacrifice from its birth to the age of 10. On the day of the fete it is bound to a post on a sandy point at low water; as the tide rises the child may utter cries of terror, but they are of no avail, as it is abandoned to the waves and the sharks soon arrive to finish its agony and thus permit it to enter into heaven.*

Humanity's fascination with sharks sometimes leads us in inexplicable directions with our beliefs and myths.

Several types of sharks are known for their frequent human attacks. Of course, these sharks live and inhabit areas where people swim, dive, vacation, and frolic in the ocean, so it's usually a case of people invading their seascape, and the shark has a relatively easy target. The triple threat, or unholy trinity, of great white, bull shark, and tiger shark are the most common culprits of attacks on humans, according to research compiled through decades; a close fourth is the oceanic whitetip. They each belong to the family known as requiem sharks and are written about in some detail in this book. In many cases, the shipwrecks covered here happened in the deep ocean, the prime hunting ground of oceanic whitetips.

The great white (of *Jaws* fame) is the reigning horror of the seas, known for brutal attacks on its prey, be they humans, seals, or any other warm-blooded creature. The tiger shark is another of the maritime marauders, also responsible for the most attacks on humans. Rounding out the list is the bull shark.

The Florida Museum of Natural History categorizes "unprovoked shark attacks" on swimmers and waders; surface recreationalists, entering or exiting water; and divers. According to information from the Florida Museum of Natural History's International Shark Attack File at the University of Florida, in 2015 surfers and others engaged in board sports were most often involved with shark attacks. "Surfers have been the most-affected user group in recent decades, the probable result of the large amount of time spent by people engaged in a provocative activity (kicking of feet, splashing of hands, and 'wipeouts') in an area commonly frequented by sharks, the surf zone," museum documents say. The museum's history is as follows: "Established in 1958, [the museum] is administered . . . under the auspices of the American Elasmobranch Society, the world's foremost international organization of scientists studying sharks, skates, and rays." They recommend, if attacked, for the victim to react immediately: "If one is attacked by a shark, we advise a proactive response. Hitting a shark on the nose, ideally with an inanimate object, usually results in the shark temporarily curtailing its attack. One should try to get out of the water at this time." They say that blows to the shark's snout may dissuade the attack, and if the shark bites, go for the sensitive gills and eyes. "One should not act passively if under attack as sharks respect size and power."

The organization administers a database, studied by biological researchers, with more than 5,700 individual investigations from the mid-1500s to today. The ISAF curator is George H. Burgess; you can search for the group on Facebook and Twitter. (Responding to an email from me asking for comments about shark attacks, Mr. Burgess replied, "No offense intended, but I really have no interest in involvement with projects that sensationalize sharks.")

# A MAN EATING SHARK

The Story a Mississippi River Pilot
Tells of His Own Seeing.

Here's a story about a man overboard, printed in the *Burlington Free Press* (Vermont), from the *New Orleans Times-Democrat*, on August 29, 1900:

*Will a shark bite a living human being? The question has been debated hundreds of times and came up for discussion among a little party at a suburban resort. "In spite of the current legend," said one of the group, "I don't believe sharks will attack a living person. I have spent my life near the sea and have heard a hundred stories of swimmers being killed or bitten by the monsters, but all the tales were either at second hand or were so vague they would never have passed for evidence in court."*

*"Well, sit," said another of the party, "I believe sharks do kill men, and I have the best of reasons for my belief. I witnessed such a tragedy with my own eyes." The speaker was Captain McLaughlin, one of the oldest and best known bar pilots in the Mississippi river service.*

"It happened 21 years ago," said the captain when pressed for details, "but the circumstances are as distinct in my mind as if it had occurred only yesterday. I was out looking for ships with my partner, Captain Tom Wilson, and the usual crew, and about 12 miles off South Pass we sighted a large sailing vessel which proved to be the Zephyr, from Bath, in charge of Captain Switzer. There was a rival pilot boat nearby, and we both made a rush for the ship to get the job of taking her in.

"Our party was nearest, and Captain Wilson and two sailors put off in a small boat to go aboard, but in their hurry they made a miscalculation and were struck by the bow and capsized. It all happened in a flash, but Wilson and one of the sailors were lucky enough to get hold of the overturned boat and hang on. The other sailor was thrown some distance away into the water.

"He was a big, brawny, six foot Swede named Gus Ericsson, and when we saw him come up, one of the crew tossed him a circular life buoy, which he seized almost immediately. The buoy was amply sufficient to sustain him, and he put his arms across it and held himself out of the water fully breast high. We had another small boat and started at once to pick up the three men, making for Ericsson first.

"When we were less than 100 feet away, I saw a gigantic tiger shark rise and start toward him, and at the next instant the poor fellow shot down out of sight, life buoy and all, like a man going through a trap. We were so horrified that we simply sat still and stared, and what seemed to be two or three minutes elapsed. Then the life buoy suddenly *appeared. It must have risen from a great depth, because it bounded at least four feet into the air and fell back with a splash. Of Ericsson we never saw a trace. He went into that shark's jaw as surely as two and two make four.*

*"We rescued the other men all right," said Captain McLaughlin in conclusion, "and Captain Wilson is still alive to bear out what I say. That, gentlemen, is my reason for believing that sharks will attack human beings. However, if anyone can tell me what became of Ericsson, I am open to conviction."*

*—New Orleans Times-Democrat*

What would you do? Hold still, hoping the shark swims by, watching you with its cold eyes but unthreatened? Do you play dead to save your life, perhaps? In the water, you can't run—even if your feet touch the bottom, you know you can't get away. What would you do during an attack? Passively hope the shark just keeps swimming, never imagining that its senses are tweaked and zeroing in on you as the prime target for a meal or for elimination from the threat pool? Maybe your amygdala is screaming, "red alert," but your brain's frontal lobe overrules and calmly and rationally tells you that ignoring the shark will make it go away—like the boogeyman in a nightmare. But you also know you're powerless in the face of this terrifically awful powerful beast whose only purpose is to kill and eat. You know you're an alien in its aquatic world and you might be eliminated.

A question to ponder: *In the throes of a shark attack, how would you respond?*

## The National Marine Fisheries Service National Oceanic and Atmospheric Administration (NOAA) addresses the question: What Causes Sharks to Attacks Humans?

<u>This interesting and helpful information is from www.nmfs. noaa.gov:</u> Sharks do not normally hunt humans. When they do attack a human, it is usually a case of mistaken identity. Sharks sometimes mistake humans for their natural prey, such as fish or a marine mammal or sea turtle, and most often will release the person after the first bite. The majority of shark bites are "hit-and-run" attacks by smaller species, such as blacktip and spinner sharks. They mistake thrashing arms or dangling feet as prey, dart in, bite, and let go when they realize it's not a fish. The "big three" species—bull, tiger, and great white sharks—are big enough to do a lot of damage to a human and must be treated with respect and caution.

<u>Is There an Increase in the Number of Sharks and Attacks?</u> In 2001, there were seventy-six recorded unprovoked shark attacks in the US, versus eighty-six in 2000. According to the International Shark Attack File, the numbers of shark bites from year-to-year seem to be directly associated with increased numbers of humans swimming, diving, and surfing in the ocean. Some shark populations have been on the decline since the mid-1980s, when the commercial fishery for sharks became a booming industry. Current regulations are working to reverse the trend of declining shark populations in the

US, although some species are still depleted, and to maintain the shark populations that are healthy.

What is NOAA Fisheries' Role With Sharks? The National Marine Fisheries Service (NOAA Fisheries) manages the commercial and recreational shark fisheries in the Atlantic Ocean, including the Caribbean Sea and the Gulf of Mexico. In the Pacific Ocean, NOAA Fisheries works with regional fishery management councils and is developing shark management measures. The agency is mandated by Congress under the Magnuson-Stevens Fishery Conservation and Management Act to conduct stock assessments, monitor the species abundance of sharks, and implement fishery regulations that maximize the benefits of sharks as a resource for humans while also ensuring that we do not deplete shark populations. The United States began regulating shark fisheries in 1993. A new Fishery Management Plan that included sharks, swordfish, and tunas went into effect in 1999, and sharks have been regulated under a catch limit and quota system ever since.

Why Should We Protect Sharks? Sharks are simply awesome creatures whose biology has remained virtually unchanged for millions of years. Just as humans strive to protect other living creatures from becoming threatened or endangered, it is our duty as stewards of the Earth to protect all ocean life, including sharks. As top predators in the sea, sharks provide a valuable balance to the marine ecosystem. Humans are one of only a few species that prey on sharks (killer whales and other sharks are others), killing over a hundred million per year. We must support and abide by fishing regulations that were put into place to ensure that sharks will thrive in the ocean for millions of years to come.

How Common Are Shark Attacks? How Do I Minimize the Risk of Being Bitten by a Shark? More people are killed each year by electrocution by Christmas tree lights than by shark attacks. Think about the things you would do to minimize your family's risk of being harmed by Christmas tree lights. You'd unplug the lights at night and never leave them unattended. You'd keep your tree moist to prevent a fire. Maybe you'd educate your children about the potential of electric shock if they improperly plugged in the lights. Similarly, you can take precautions that minimize your risk of encountering a shark when visiting the beach this summer:

- Always stay in groups, since sharks are more likely to attack an individual. Do not wander too far from shore— this isolates you and decreases your chance of being rescued.
- Avoid being in the water early in the morning and during darkness or twilight hours when sharks are most active and searching for food.
- Do not enter the water if bleeding.
- Avoid wearing shiny jewelry because the reflected light resembles the sheen of fish scales.
- Avoid waters being used by sport or commercial fisherman, especially if there are signs of bait fishes or feeding activity. Diving seabirds are good indicators of such action.

- Use extra caution when waters are murky and avoid bright colored clothing—sharks see contrast particularly well. Refrain from excess splashing.
- Exercise caution when occupying the area between sandbars or near steep drop-offs—these are favorite hangouts for sharks.
- Do not enter the water if sharks are known to be present and evacuate the water if sharks are seen while there. And do not approach a shark if you see one.
- Between the months of May to September, restrict your ocean swimming from 9 to 5.

# WHAT *I-58* CAUSED

*—Story by Joseph B. Healy*

The sinking of the USS *Indianapolis* by the
Japanese submarine *I-58*, July 30-31, 1945

## THE USS *INDIANAPOLIS*

Introduction by Stephen H. Foreman

*July 26, 1945—The mission of the USS* Indianapolis *was
to deliver components for the first operational atom bomb to
the South Pacific island of Tinian. Mission accomplished.
She was then ordered to join the battleship USS* Idaho *in the
Leyte Gulf in the Philippines to prepare for the invasion of
Japan. There was no escort ship, as was usual procedure.
The skipper of the* Indianapolis, *Captain Charles B. McVay
III, requested a destroyer escort—in an apparent blunder
of Naval command, this was denied—but was assured the
route was safe. It was not.*

*July 30, 1945—Twenty-year-old Corporal Harrell was
a Marine on deck watch at midnight, July 29, 1945.
After his watch he went below decks to his berth; however,*

because it was so hot down there, he returned topside and slept on the deck under the barrels of forward turret #1. The Indianapolis was between Guam and the Leyte Gulf when the cruiser was stopped by two torpedoes fired from a Japanese submarine. According to Corporal Harrell, "Metal groaned and twisted, water churned and rose, and men scrambled and screamed." She sank in twelve minutes. Of the 1,196 servicemen aboard, 900 men went into the water—317 would survive. Only a few rafts made it into the water with them. Life jackets were standard issue kapok. It was pitch black. The men were floating in the middle of the ocean. They could only hope for rescue as soon as possible. Some were severely burned from explosions, some had broken bones and cuts, most were covered with fuel oil discharged in the water as the ship broke into pieces. The next four days were a gruesome and harrowing tale of survival. More men died here than in all the disasters in naval history.

Shark attacks began at sunrise. White tip sharks. Imagine the fear of these men, the abject panic, as they observed the sharks circling them before they attacked, and then when the sharks began picking off men, dragging them under, tearing off their arms and legs, biting them in two. To be hanging onto a life raft or treading water, and to see the man not five feet away bitten, killed, where he was now a patch of blood red water. Another survivor, Willie James, said, "The sharks were around, hundreds of them. Everything would be quiet, and then you'd hear somebody scream, and you knew a shark had got them."

It was five days before they were rescued, five days exposed to the elements without fresh water, five days filled with continuous, deadly shark attacks, days of hunger and dehydration. Tongues swelled, lips split open. When the ocean water dried out from the sun it left salt caked on their faces and in their eyes. Terrible thirst and dehydration set in, yet to drink salt water was to drink poison. Some of the men became so desperate they couldn't hold out and drank the salty water, anyway. "It took only about an hour," Harrell said, "before they began hallucinating"—hallucinations terrifying and final.

This was a tragedy that did not need to happen. Before the Indianapolis set out naval intelligence broke the Japanese code and learned that two enemy submarines were in the area. Captain McVay was not told that a Japanese submarine had recently sunk a destroyer escort as well as a battleship in the very waters to which the Indianapolis was headed, although the information on the attacks was received by naval intelligence. As the Indianapolis sank, three distress signals were sent out and received at three separate locations but were ignored because they were thought to be a Japanese trick to lure American vessels into the area. Although Captain McVay was a beloved and respected commander, he was court-martialed and blamed for the disaster.

More—much more—follows in our account of one of history's worst shark attacks.

Eternal Father, strong to save,

Whose arm hath bound the restless wave,

Who bidd'st the mighty ocean deep

Its own appointed limits keep;

Oh, hear us when we cry to Thee,

For those in peril on the sea!

O Christ! Whose voice the waters heard

And hushed their raging at Thy word,

Who walkedst on the foaming deep,

And calm amidst its rage didst sleep;

Oh, hear us when we cry to Thee,

For those in peril on the sea!

*—Words from "Eternal Father, Strong to Save,"*
*considered the hymn of the US Navy*

Stoic and mindful in the garden of Umenomiyn Shrine in Kyoto, Japan. Staring at the staid red *torii* marking this sacred space as a welcoming gateway, Mochitsura Hashimoto allows his thoughts to float up and circulate in his mind but is not moved to any action. He gently guides these thoughts and regulates himself: Sit calmly and let the thoughts come, let them visit, and let them leave. Welcome their visit and bid them good-bye: They tell us, remind us, enchant us, entertain us, but they also must leave us.

Vivid and powerful thoughts are embedded in his memory, creating inner voices that are his constant companions, whether he likes what they tell him or not. He doesn't dislike knowing and remembering—he simply does, the thoughts come and he allows them to visit and knows that this was his life and he is thankful for having such a full life. These reflections are part of his meditation.

Mochitsura Hashimoto is eighty-nine years old and living comfortably at Umenomiyn, fulfilling the expectation of his father that he would continue the family heritage as a *Shinshoku* or Shinto priest, as the father had been. His father died before Hashimoto would commit to the Shinto path—many decades before and indeed for Hashimoto what became an unimaginable lifetime later, long after the war and the devastation and destruction, the pain and eventual rebirth. Long after Hashimoto gave the order to sink an enemy ship during World War II. That ship was the USS *Indianapolis,* which had just delivered the components for the atom bomb, nicknamed Little Boy, dropped on the Japanese city of Hiroshima soon after.

Hashimoto listened to his father and as a boy had considered the Shinto path. His father prayed and meditated, and ultimately relented, directing his young son otherwise, encouraging him to enlist in the military. Previously, the circumstances of Hashimoto's youth—seeing the financial strictures of his father's meager state-sponsored stipend for serving as a Shinto priest, and then the war—led Hashimoto's interests toward other horizons, too. He had witnessed the difficulties his father had raising a family on the meager pay of a priest and decided he would not struggle that way. Of course, he wanted a family and he wanted to be a man of standing who could lead his flock, but he also wanted to rise above his status and be as successful as life would allow. He joined the military as a young man, following his two older brothers into service, and studied at the Imperial Japanese Naval Academy at Eta Jima. From his simple origins in Kyoto, he found his studies thrilling—judo and athletics and engineering. He was commissioned in 1931, and by 1937, he married. His wife gave him three sons and a daughter, through the war years and beyond. This was a family of great happiness.

In the military, he was drawn to the disciplined team approach of submarine service, united by purpose and in action, but at the same time independent. He was chosen for submarine school, entered on December 1 as a Class B student, thrived, and was assigned as a torpedo officer to the boat *I-123*. Then, on board *I-24*, he was at Pearl Harbor for the attack. "As we lay concealed, we could see the . . . neon signs on Waikiki Beach. There were some dazzling lights, which we took to be searchlights . . . we could hear a radio churning out jazz music. It was close on 11 p.m. and the enemy was completely unaware of our presence," he wrote in the book *Sunk: The Story of the Japanese Submarine Fleet, 1941-1945*. Notorious now as the attack that plunged the US into World War II, Winston Churchill wrote in *The Grand Alliance*, his grand book of military strategy, asking President Truman by phone if the news were true: "It's quite true," Truman replied. "They have attacked us at Pearl Harbour. We are all in the same boat now." Churchill continues: "We had no idea that any serious losses had been inflicted on the United States Navy. They did not wail or lament that their country was at war. They wasted no words in reproach or sorrow." However, Churchill does add, "In fact, they might have been delivered from a long pain." And he continues: "How long the war would last or in what fashion it would end, no man could tell, nor did I at this moment care. Once again in our long Island history we should emerge, however mauled or mutilated, safe and victorious. We should not be wiped out."

Churchill had studied the American Civil War and knew of the grit and determination of the American people—staunch woes for a country (Japan) that believed in a divine blessing, perhaps a Shinto blessing, to win the war and dominate and prosper. "The Japanese High Command had shown the utmost skill and daring in making and executing their plans," Churchill wrote in *The Hinge of Fate*. "They started however upon a foundation which did not measure world forces in true proportion. They never comprehended the vast talent of the United States."

Hashimoto next attended advanced submarine school to prepare for service as a commander and became lieutenant commander of *I-58*. He carried *kaitens*, or manned torpedoes—suicide bombs piloted by soldiers willingly giving their lives for the Emperor, the sea's equivalent of the sky's *kamikazi*. The sub was at Iwo Jima and Okinawa, repeatedly attacked by aircraft, seldom surfacing.

Hashimoto ordered *I-58* to the Philippines, and on July 29, 1945, the sub was patrolling shipping lanes among Guam, Leyte, Palau, and Okinawa. They spied a heavy cruiser—the *Indianapolis*. The ship moved through the relative darkness of a half-moon, causing the captain of the *Indianapolis*, Captain Charles B. McVay III, to call off the evasive zigzagging pattern he had been advised to follow. The navigator shouted to Hashimoto and the *I-58* crew, "Bearing red nine-zero degrees, a possible enemy ship."

In such a moment of weighty decisions, Hashimoto was forced to adjudicate the type of ship, its course and speed, and ready his vessel for action. "There was a large forward mast. We've got her, I thought," Hashimoto wrote in his postwar book *Sunk*. "She had two turrets aft and a large tower mast. I took her to be a large *Idaho*-class battleship."

He had tentatively thought of sending two suicide *kaitens* with six conventional torpedoes. He sent the six conventional, holding the *kaitens* for a follow up to the salvo if needed, much to the bitter disappointment of the *kaitens*. He wrote of the torpedo launch:

> . . . [T]here rose columns of water to be followed immediately by flashes of bright red flame. Then another column of water arose from alongside the Number 2 turret and seemed to envelop the whole ship—"A hit, a hit!" I shouted as each torpedo struck home, and the crew danced for joy.

Looking through the periscope at the blasts, Hashimoto knew at least three explosions occurred. More explosions reverberated over and through the sea, likely internal to the *Indianapolis*. The commander was quite sure no ship could escape the resulting horrible fate—the cruiser could not move; it would sink; men would die—many men would die. Hashimoto did not mourn the loss of enemy life, knowing he also was expendable and his responsibility rested with preserving the lives of his crew. Maybe that was his greatest responsibility? He wanted victory for himself and Japan; he wanted to defeat the enemy. But he wanted his crew to survive. The *I-58*, picking up the underwater detector signal from the *Indianapolis*, dove to evade the signal and to reload, preparing for the likelihood further commands were needed. Hashimoto acknowledged the pleas of the *kaiten* crew to send them on their final mission, but he didn't want to spend their lives if the enemy ship was already sinking. Why sacrifice men when the job was already done, the enemy vanquished? When the submarine reached periscope depth, nothing could be seen. There was no flotsam—no evidence of a sunken ship. No signal or sign of hundreds of bodies afloat in the water, no wreckage. Nothing. They would eat, feast on the best they had—rice, beans, and tinned boiled eels and corned beef, a gratifying meal. A message was sent to Tokyo: "RELEASED SIX TORPEDOES AND SCORED THREE AT BATTLESHIP OF IDAHO CLASS— DEFINITELY SANK IT."

What could not be seen through a periscope, only experienced in the hopeless dark of a nightmare, was the salt water surging overhead, the waves swelling, the piercing screams swollen with terror, guttural cries for the mercy of an angry or vengeful God, invoking their God's benevolence to provide forgiveness and salvation. "Oh, my God, I'm heartily sorry, for having offended thee!" Brutal shouts, profane cries; pitiless begging and pleading and sobbing. Hashimoto saw none of this, nor the float path of the dead extending out from where the cruiser went down.

(That path would stretch out twenty-five miles or more within days.) Those dead, dying, or fighting for survival were in a nightmare seascape of human suffering. One man grabbed another by the shoulders, but that sailor was gone, burned horribly beyond recognition. Another was shouting to watch out for the Japanese airplane, his mind warped from ingesting salt water, and growing hypothermia, shock, and fatigue. After some time in the water, clarity came to some, and they clutched anything that floated, pulling their head, shoulders, and, if possible, their torsos clear of the water. Some rolled their bodies prone on a piece of wood or metal or a kapok cushion. Cries for God, and mommy, too.

As reported in the book *Fatal Voyage: The Sinking of the USS* Indianapolis by Dan Kurzman: "As soon as Captain McVay returned fully dressed to the bridge from his emergency cabin, Commander Flynn confronted him. 'We are definitely going down,' he said, 'and I suggest we abandon ship.'" This time-proven military leader, Captain McVay, had just heard words he never imagined he would hear, ever: and next he spoke words beyond his own comprehension: "Abandon ship!"

It was so sudden. The ship was gone—disappeared, disintegrated from view. Sunk in fewer than fifteen minutes.

With clarity came awareness that the men were not alone floating in the water. Shadows moved past their groups, as they had melded together into pods of five, twenty, fifty, as many as 170, the approximately 900 who survived the sinking now grouping together when possible and when their flagging strength allowed. Safety in numbers, of course, but also a feeling of security knowing they were not alone in what they faced. A shadow passed a weary man while another, floating in the current face down and clearly dead, suddenly convulsed seemingly in a spasm. No, the ocean itself surged and erupted, the limp body jerking under the surface, like an electric shock, then the body surfacing briefly, jerking down again. The body vanished in the dusky light, as dawn was now coming and forms were taking shape. A chorus of shouts became one high-pitched scream, gurgling out to a startling silence. The sharks had arrived and food was abundant. They were feeding, frenzying amid some groups of men, bodies disappearing, not all dead when the attack began but lost in a swirl of blood and foaming water.

Hours passed, the unforgiving sun beat its rays on the men, the 110-degree heat intolerable. The surface was covered with a fuel-oil skim, a choking thirst overwhelming the parched sailors. More shouting and splashing, then screams, then quiet. Other voices were threatening, shouts accusatory, violent splashes not sharks but sailors turning on one another, the insanity of dehydration and exposure erupting in a cacophony of violence. This attracted the sharks, of course, who quieted the thrashing men, one at a time, picking them off the surface. Some were taken in mid-thrust as they splashed at another sailor, shouting insane invectives—silenced.

One group that started as seventeen men in a loose circle was now reduced to ten, their twenty arms clutching wooden boxes and boards. Together they splashed the surface and kicked as two sharks circled, though one shark turned swiftly, seemingly angrily, and swiped a man's legs, pulling the sailor's torso under, the man spitting blood when he resurfaced, never shouting just gasping and spitting and finally lying back in the water, his head jerking down out of view. Another man lost.

What was coming next—a Japanese sub surfacing to strafe the men with machine-gun fire? Or would planes be dispatched to locate the wreckage and survivors? The likeliest scenario was the return of the submarine and certain death. *Well,* a sailor thought, *get on with that! Don't let us drown here or lose our minds to kill one another— the sad irony of surviving a sinking only to lash out and kill beyond awareness because you were driven by madness caused by thirst and hunger and pain. What was that? Another shadow? Damn sharks—take me! No, don't you come near, I'll kill you with my bare hands you rotten bastard! I'll gouge out your lifeless black eyes and drink your blood! My God, that's it—come near and I'll eat you and drink your blood!*

The shark turned and dove, quickly angling up to attack from below. The sailor could see it clearly, could see its teeth bared. He spun around and tried to swim, fiercely thrashing his arms, gulping seawater, losing his breath, feeling searing pain in his legs, knowing there would be no fight, screaming, *"Nooooo!"* His last memory was watching the surface rise up as a second shark collided into this chest, knocking his torso backward while his legs were being pulled down, the pain so bad it brought euphoria, with him ascending now ever higher till he was above the water's surface and flying now toward an island and safety, heading home. He missed home. How he missed home.

How long had it been? Why had no one come? Neither airplanes nor rescue ships nor an enemy sub, nothing had materialized—the men in the water were abandoned. Alone in the water, how could they possibly survive? There was no way to survive—no food, barely any fresh water stowed in the rafts, no painkillers, no medicine, no sleep. Only pain and deprivation beyond the limits of the human body and mind. And there were sharks prowling and patrolling the edges of the groups of men.

*The human mind has great capacity. We can transport ourselves to calming nocturnal moments of wonder and joy. We can experience alpha and omega in the* OM. *We can allow our minds to be absorbed into the great wonder of existence, beyond our awareness to a deeper place of calm. To reach an absolute calm, a tranquility, a peace. We can rest and believe that some life force will repel all evil. Even the sharks aren't evil, though they look it—they simply want to go on living and need to eat. You are not shark food—until they decide that they must eat and you are there. This is what Shinto has taught me,* Hashimoto thought as he woke from his horror dream. *I will live in that temple in Kyoto, one day I will help. But not today.*

He rose from his bunk and went to the head to piss. His stomach churned from the rotten onions he had eaten earlier, simply to show his crew they were edible. *Whosoever died on the American ship deserved to perish, because they wanted to kill me and my crew and all of Japan,* he thought. *Still, they are not* they—*I can't hold them as a collective in my mind,* he corrected his thoughts,

*I know many individuals were alone with their life forces and didn't have any decision about dying. That man had to die. I did not want individuals to die. I want Japan to live, and we will live. However, there is no beauty here where I sit below the water in a steel tube that is a submarine, but I remember beauty and the gardens of my youth and the delicate flow of nature. I love* Kami. *I respect* Kami. *I know I am respected by my crew, although I also know they see me as old and therefore wrongheaded, but I will maintain respect for life throughout my own life. Down here, I feel only my responsibility to the Emperor. I am not in touch with* Kami, *but I will be one day. That day will come when I am ready*, Hashimoto told himself.

Coxswain Louis Harold Erwin told the History Project of the Library of Congress:

> I'd just come off the eight to 12 watch, stood—stood my watches on the five-inch gun. And I just got in my hammock, and this big blast hit. And the ship give a big lisp, so we—I was out of my sack, and I said this ship's going down. And we all carried these big knives on the side, and our life jackets was—he kapoks was put in big bags up there and tied along the railings and different things, and we start—took our knife out, cut the kapoks down and started passing them out. So we never did hear abandon ship. All the communication was knocked out. And we kept seeing a group of people in the water, so we said we better hit it or we're going down with the ship. So I run down the side. A few more of my buddies run down the side and we dove in. And the

first night—this happened about 12 minutes past midnight. All the rest of us, we heaved and lost everything we had on our stomach drinking that salt water, swallowing that saltwater and oil. And the next morning, why, of course, I'd swam just as far as I could, and I looked around, and I just saw the tail end of the USS *Indianapolis* going straight down.

[T]hat first night was real light. That's when you swam as fast as you can to get away from the ship, I could look back and see the fantail of the ship going down, and you could see that. But many of the nights it would get awful dark. And of course, when the sharks attacked, that would be—that would be earlier, feeding time, sort of time, and you could hear screaming at night or different things when they was getting someone. Sharks would swim within five to six feet from you, knowing all the time that they could get you.

And we tried to all gather in the water and keep ourself (sic) together where the sharks wouldn't get us, but the group I was in, about 250 to 300, after first day we'd lose a few, second day a few, just kept on. About the third day, why, I looked around, just about all of them was gone.

The Oceanic whitetip shark is a requiem shark, meaning it produces live offspring, is migratory, and lives in warm, deep oceans (as opposed to shallower, inshore areas).

According to the National Oceanic and Atmospheric Administration, "Oceanic whitetip sharks are moderately large sharks with a global distribution. This stocky shark is easily distinguished from other sharks by its unmistakable whitish-tipped first dorsal, pectoral, pelvic, and caudal fins. It has a large rounded first dorsal fin and very long and wide paddle-like pectoral fins with a short, bluntly rounded nose and small circular eyes. The oceanic whitetip shark is a pelagic species that lives near the surface in warm waters (usually over 20 degrees Celsius) in the open ocean, usually well offshore. [They] are found worldwide in warm tropical and subtropical waters between 20° North and 20° South latitude, but can be found up to about 30° North and South latitude during seasonal movements to higher latitudes in the summer months."

They eat whatever they find. Survivor reports indicate that tiger sharks likely joined the feeding frenzy.

In the water in the straights of Leyte, hundreds of men continued their struggle. Most knew intuitively from their inborn survival instincts that they needed to belong to a group and were eager to be a part of the whole. Every once in a while, an oil-faced sailor would erupt with epithets leading to violent action; every once in a while, a man died at the hands of a delirious shipmate. In the throes of delirium, once, the sailor was taken down by a shark, the thrashing and convulsing an attractant to the apex predator underneath. No part of that man resurfaced; he was consumed below. The other sailors were glad *not* to see remains—they couldn't bear another reminder of eminent death surrounding them.

How their destiny now was intertwined with the reality of death. *Growing up on a farm in Iowa, walking the fields through harrow furrows, kicking at cornstalks trying to force pheasants into the air to bag dinner for his family, his father a wretched drunk passed out behind the barn, the boy now the provider for his two younger sisters and his mother overwrought with nerves, barely able to keep the cast-iron pan steady on the cook stove so terrible were her tremors (maybe she was drinking too?), he felt the familiar isolation now of being in a vast sea with no direction and little hope. What did it mean to survive, anymore; he had been trying to survive his entire life and was responsible for the survival of others, as he felt he now was too. He was an ensign, the lowest rank of officer but a commissioned officer nonetheless, making his own way only to be swept up in the war machine and now dropped, discarded, into the south sea like so many unwanted bundles and scraps. NO! He would not submit to that morbid thinking, he would persevere and prevail and survive. He would return to Iowa with his Naval commendation or Purple Heart for the wound he received as the ship went down. The wound . . . his leg ached so, and it made him more tired than hungry or thirsty. If only he closed his eyes for only a moment. He splashed water on his head and tried to wipe the oil from his face, grinding his knuckles into his eye sockets. He could see clearer; he would sleep better; he would now close his eyes and let sleep drift over him. It felt so good and easy. He would wake up and be somewhere else. He would sleep and float away.*

Survivor Giles McCoy was interviewed for the Library of Congress Veterans History Project on November 14, 2002; he talked about being in the water after the USS *Indianapolis* went down:

> [T]he weather was rough, windy and a there was a lot of waves; there was so much oil (fuel oil) on the surface that I swallowed a whole bunch. When you swallowed [fuel oil] it made you sicker than the devil. You just vomited all the time until your insides felt like they were coming out; it just got so bad. Anyhow, I was floating out there and I came across another group of men and got with them. One of them was a bosun's mate that was a good friend of mine that I had done duty with; his name was Gene Morgan. Gene was straddling a five inch powder can, he didn't have a life jacket. I got up to him and he recognized me. I said, "Gene, you can't survive on that powder can. We don't know how long we are going to be out here." He said, "Well, that's all I got!" We had to wait until a dead body floated by with a life jacket on; I took it off the body and said a prayer to it and let it go. I put [the jacket] on him and then I left him.
>
> A group of fellows had a piece of a raft that had come up and it was damaged from the second torpedo, I guess. The whole front of the raft was all blown off. If you look at that poster you can see the raft [points at detail of poster]. That is the only thing that was out there; they didn't have a bottom

to it. There was already a bunch of guys that were holding onto what was left of it. I got there and Gene didn't want to go. He said, "No, they will pick us up tomorrow." I said, "It may not be tomorrow." And it wasn't.

Anyhow, I got there and (I had swallowed a lot of oil and was sick) one of the guys got hold of me and helped me get my life jacket tied on and helped me get through my vomiting and all. Then we started, and we wound up with seventeen of us all together.

There was no bottom in the raft, it was just a ring, a balsam ring; you had to have a life jacket otherwise you would drown. Most of us were hanging on the outside and then when daylight came the sharks came. We had sharks everywhere. The first couple of days there was probably a hundred sharks around us all the time. A couple of guys got hit by sharks and got taken down. We did everything wrong. We kicked our feet and tried to get them up out of the water and we climbed on top of one another because we knew (sharks) would come underneath you and come up after you. Even where the raft was damaged, blown apart (the rope netting that was inside the raft) they would come up and would swim with their head into the raft. I kicked seven of them in my days, kicking their nose out of that raft. I never had any training on sharks, I was just told you had to try to keep away from them. If you got hit by one of them you were pretty well going to die, because it would attract a lot more sharks. I found out that if you hit them in the eye, kicked them in the eye ball, it really hurt them and they would leave you alone. You could see thirty, forty, fifty feet down into the water and you see the thrash back and forth after you hit them. They just couldn't stand that, and they left you alone; the sharks that you kicked in the eye, they didn't come back. The other ones would come back, but not those.

Anyway, we went through the time and the days went on. I kept trying to encourage everybody that within forty-eight hours they would be picking us up out of the water. I kept telling a lot of the guys that were wanting to give up (guys started giving up quick) I would tell them, "You can't die, you got to stay alive, you've got a family at home, hang in there." I said, "I don't have a family, I just have a wonderful mother and a bunch of sisters and my father and I want to stay alive for them if I can. Some of you guys got families." So we tried to help one another and as the days went on and nobody showed up we realized we were not going to be picked up out of the water. We were going to eventually die and I didn't want to end up in the belly of some shark and neither did the other guys.

Some of the men were delirious and were hallucinating. "A lot of them had [that] bad. They were seeing things and they were hallucinating. It was easier to die than it was to stay alive; to stay alive you had to work at it, but to die all you had to do was quit, just give up. We kept trying to encourage the guys. They wanted the easy way out, they had all [the suffering] they wanted. We kept a lot of them alive.

How many hours had passed in this cataclysmic world? One of the men thought, *Could this be a version of hell—suffering, pain, torture to an unreal degree. Did we all die when the boat went down and now this was hell? The world was upside down and this was the penury and pain he'd learned from the nuns as a boy at St. Margaret's Catholic School. When he prayed on the rosary, this was the hell he envisioned—never ending, the forevermore of suffering. Could it, would it end? If he dove into the water, the sharks would kill him. If he stayed in the raft, the sun would bake his skin, and he had nothing to eat or drink so that would kill him, too. He had another thought: If this was hell, how could he die—again. Maybe the sharks wouldn't eat him but only chew on him, he wouldn't die but be in constant pain. Better to stay in the raft. Although the man across the raft was giving him looks, angry looks—he could spring forward at any moment and begin the unending torment, the same as the sharks. Mayhem, delirium, brutality, violence—he did not want to die at the hands of a fellow American sailor. But the sharks. Maybe the sharks were the answer. This was purgatory and the sharks would provide an end to the suffering and salvation would be next.*

*What if I have to endure more suffering to reach salvation, and then the Kingdom of God will be mine.*

*He tried to remember his catechisms . . . but he could not. Saint Peter, Saint Bartholomew, Jesus on the Cross. That was suffering. My God, my God, why have you forsaken us?*

*He thought of Jesus on the cross and in Golgotha. He said an Our Father and rolled over the side of the raft, starting to swim and then splashing maniacally. End the torment, end the pain, end of the suffering. The man was lifted by a wave and as he slid into the trough, as if summoned by his will, he saw the shark rising toward him. He undid the kapok life jacket, which had rubbed his neck raw, and began to splash harder, flailing his arms on the oily surface and kicking below spasmodically, and was ripped underwater in a swirl of bubbles on then a cloud of blood. Salvation had found him, at last.*

"Do you hear that," a yeoman tried to shout to the oil-covered sailor across the raft from him. The weak man croaked, *What?* "A plane, an engine, there's somebody coming," the first man answered. As soon as he said it, his hopes deflated, remembering the other planes that had flown overhead and in spite of their thrashing and waving, had kept on flying. Once, a man swam out into the water and was about twenty yards from the raft, riding a wave, when he disappeared like he had never been there and was only an apparition. A shark took him. Still, the yeoman girded his hopes: "I hear an engine and it's getting louder!" Indeed, he heard the Ventura under the command of Lieutenant Wilbur C. Gwinn. The date was August 2, 1945.

As he was peering through the gunner hatch at the plane's tail section, while trying to fix a navigation antenna on the fuselage of the plane, Wilber Gwinn had spotted an oil slick on the water's surface. He ordered his copilot to dive toward the slick. He rushed to the cockpit, and as they got closer he could discern black dots in the water—maybe Japanese soldiers, adrift after their sub went down? "There're people down there," he shouted, "gotta be Japanese!" The plane descended and he could now see splashes and arms waving. Indeed, people down there, in the middle of the Gulf of Philippines, floating and splashing in an oil slick. At a glance, he also could see shadowy forms outside the cluster of men, the black dots gathered in a tight school with ominous shadows on the outskirts; he happened to register a shadow rise up into one splotch of splashes and then there was calm. He immediately understood—it was a shark attack. *The goddamn sharks are feasting!* he thought, and a shiver went through him. Drowning or burning alive, he could imagine that in war, but not being a sitting duck for sharks underneath, human fodder floating helpless until a stupid, carnivorous creature decided you were food and made a meal of you, gnashing and slashing and tearing away your flesh and limbs. The horror of that—truly horrific, almost unimaginable and unspeakable. This flashed through his mind and he soared overhead. Poor bastards, whoever you are.

The rescue operation began. Gwinn made a water landing, and as another plane circled overhead, radioing directions, Gwinn kept his engines roaring and started to skim the surface, picking up survivors. The door was opened on *The Playmate 2,* and a soldier leaned out and pulled men toward the rope ladder, once grabbing a man's shoulders to hoist him, only to have the lifeless body—half a body, the torso and legs gone—bob and roll away from his grasp. The sailor was half eaten! He would never forget the emptiness of the face, no pain, no surprise, just an empty stare. Jesus, he had no legs, half his body was gone!

*"Half his body was gone,"* he repeated aloud to himself.

Others in the water were whole, at least physiologically, though in aspect they seemed more like newborn birds, wizened and withered and shriveled and hairless (doctors later explained that prolonged salt water exposure caused hair to fall out), trembling like hatchlings. Some sailors made a faint *peeping* or *chirping*, too, calling out in distress. More than their look, the *peeping* pissed off a flyer more than anything—why did they have to cry out like that? What the hell were they trying to say, anyway?

The rescuers learned this was the crew of the USS *Indianapolis*, one saved sailor told them that. They had gone down quickly after explosions happened, must've been a mine. Or a torpedo, the flyer inquired? Maybe, coulda been. I dunno, it happened so fast, all of a sudden we was going down. "I saw two of my friends disappear, right next to me, it was the sharks. We was one the side of a raft, we couldn't get up into it, and we was just hanging on. And then they's were gone. No screamin', no fightin', just gone."

Louis Harold Erwin told the Veterans History Project:

> The sad part about losing this ship and the crew is they did not start looking for us and we spent so long time in the water. If Leyte in the Philippines would have notified that the ship hadn't arrived,

then maybe they'd have started looking for it. We were very lucky that a bomber come over, piloted by Lieutenant Gwynn, Chuck Gwynn, and he accidentally just spotted the heads. He was flying low. He just accidentally spotted the heads out of the water down there, and he went back to—or sent someone back to look. His radio antenna was broke, and they happened to see these heads bobbing out of the water. And he made a run over us, and he was about to drop a bomb on us, and he looked and saw us waving and different things. So he spotted us and radioed back to Peleliu, which then that's when the PBY come.

And so on the fourth day, why, [after] this plane spotted us, and this PBY piloted by Adrian Marks, Lieutenant Adrian Marks, when he picked us up, why, he—there was only 56 of us. They was just—the reason he landed in our little ol' section there, he would see sharks attacking the men. So there was a few nets got off. The life rafts are very few and I never saw—or the group I was in, me myself never saw a life raft or a net while I was in there. We spent all of our time in a kapok life jacket. After about three days, where the people would drink that saltwater and go berserk, they'd just pull off their life jackets and go down, and you could paddle around and get you another life jacket for those kapoks will get water soaked, and they begin to give out on you. And the worst part about a kapok is when you're in one and you're in the water, why, your head's back like this [indicating], and when a wave come, it just comes over and covers you. After about the second—first and second day, I just scooted mine off my shoulders and come back and sort of sit in it and brought my head up out of the water where I wouldn't get all that stuff. And after they come—the ships started arriving the next morning—I was picked up around 5:00, 5:30 that afternoon. We spent—about 2:00 or 3:00 o'clock in the morning they started coming, when the ships started coming to rescue us, they took us off the wing of the plane and put us on a *Cecil J. Doyle*. Of course, there was all kinds of rescue ships there by then, and each one had different— picked up different groups.

Long after nightfall, every space inside the plane, and on its wings, was full—fifty-six men out of the sea, resting dry in the fuselage of the plane. Gwinn feared that the plane might not stay afloat; it didn't help that the hallucinating sailors kept kicking holes in the sides of the plane. Sometimes, a few punches to the face made them stop, and then they would understand they'd been picked up in a plane and that this was real and that they'd survive. One man kept yelling: "Sharks'll get ya, don't keep swimming, they'll eat you. You would do it, too, you would eat me because you have to keep living. I would eat you too, motherfrigger!"

Despite the rescue effort, the death didn't end; the killing kept on. Not from the enemy, no, they had propelled armed torpedoes on a deadly course, and had now fled. Those being killed suffered from nature swimming below. They were being eaten, in wholes or in halves. Unimaginable that a thinking, feeling mind—*Homo sapiens sapiens*, which means he who knows that he knows—could have its parts eaten by an animal, devoured and digested. Those devoured whole were never seen again; they disappeared. The halves seemed the worst: alive and feeling their legs ripped away, bleeding out into the ocean, and then darkness. Some men willfully ending their lives, choosing to swim away from the survivors in the rafts, some delirious and imagining a distant island or a sunken refuge, swimming and diving and in fact offering themselves up to the sharks. With no water, hypothermia, hallucinations, saltwater sores, wounds from the torpedoing and burning debris, dehydration, delirium, loss of all hope—these deaths more euthanasia than suicide.

The PBY *Catalina* or flying boat *Playmate 2* was full with survivors, but the other rescue ships dispatched and now arriving couldn't stop the sharks from snatching down more sailors. The carnage continued. The splashing, the hollering . . . then the eerie quiet. Sometimes a torso popped back up on the surface, the kapok lifejacket lifting the remains. Now it was a matter of reclaiming the dog tags.

How many lives would be saved? A rescuer radioed: "This is all that's left of the *Indianapolis*."

A transport motored toward another raft now visible in the swells, airmen seeing a group in the spotlight, another wretched bunch huddled in a raft. A group of nine or ten. The *Ringness* proceeded and intercepted the group, pulling them aboard, the rescuing sailors deliberate and careful not to wrench or dislocate a shoulder when pulling a man to safety, so gaunt were their bodies, nearly lifeless. They learned that in this group was a man named McVay—Captain Charles Butler McVay, the commander of the *Indianapolis*. He pulled himself up the ladder and immediately collapsed, though he asked for no special care or attention. It was reported that Captain McVay wept fitfully, though remained under his own power, and was escorted to his own cabin, where he slept for hours.

Captain McVay later had a private talk with Captain Meyer of the *Ringness* about the events of the sinking. When the phrase "not zigzagging" came up, McVay at first asked that these exact words be expunged from any official record—but then reversed himself and agreed to let that descriptive language stand. These words—not zigzagging—would revisit him soon. *Zigzagging, some cockamamie French or German word*, McVay thought to himself, *that should not define my destiny*. Soon, that word—derived from *zickzack*, the meaning connoting leading in alternative directions—would become paramount in his life . . . and was there when he chose death.

One last group, nearing hopelessness as they scanned the barren surface on the fifth day adrift in the ocean, was located by the rescuers. Giles McCoy: "Then one of the rescue vessels came on the forth night. When he [the captain of the ship] was alive I told him (his name was Graham Claytor, he was the captain of the *Cecil Doyle*, which was a destroyer) . . . I told him many, many times when I had a chance to talk to him, 'You not only saved my life, because that is what gave me encouragement, but you saved many other people's lives by doing what you did.'

He did a very brave thing. When he came into the area he got the radio messages from the airplanes so he didn't wait for somebody to tell him to break off and go rescue the survivors. He just took it on his own and when he got into the area he was afraid that he was going to run over some of the survivors and kill them. He decided to put his search lights into the water and to put one up into the sky [reflecting] off the clouds to give everybody hope."

The ship came nearer and the rescuers tossed a line, but no one in the raft had the power to respond. A weight called a monkey's fist was affixed to the end of the life line, and when it landed in the raft one man clutched it—Marine Giles McCoy. The rescuers couldn't retrieve the raft with the line, so two jumped in the water without hesitating, until they saw the attendant sharks circling the raft. Two other sailors joined the first pair and made it to the raft, cutting free the heavy kapok lifejackets of the sailors and further liberating the desperate survivors. In the raft was Giles Gilbert "Doc" McCoy, a Marine sergeant who became Captain McVay's driver while still in the Philippines and later the primary catalyst for the exoneration of McVay from his inevitable court martial. He had promised God that if he was saved, he would no longer kill and he would become a doctor, and he did. In an interview, McCoy said about being in the raft: "To stay alive, you had to have a drive behind you, and to really want to be the last to die. It was easier to die than to stay alive . . . What drove me [was] my love for my mother. I didn't want to disappoint her." In another interview with the Veteran's History Project, he said, "My mother had to sign for me because I was only 17 years old when I joined the Marine Corps. I'm very fortunate to have had a mother like her."

He said about the sinking of the ship after the torpedoes hit: "I knew that I had to get away from the ship. So when I got to the keel, I just squatted down and slid on down into the water and started swimming away. And when I looked back, the ship was standing on her nose, and the propellers were still going around and men were still jumping off the fantail and many of them were hitting the propellers. And anybody after our experience that got off the ship and they were injured, they didn't have a chance to live, they were gonna die. . . . I started swimming, and I don't know how far I went down, I went down till my head felt like it was gonna blow open, and then I caught an air bubble, as some of the experts said, and I came back up in this air bubble back to the surface. And when I looked back, there was nothing left but a big old mountain of foam." The *Indianapolis*, the home to more than 1,000 men, was gone.

The atomic bomb was dropped on Hiroshima by the *Enola Gay* on August 6, at about 8:15 in the morning Japanese time. On the side of the bomb was scrawled, "This one is for the boys of the *Indianapolis*." In a city of 250,000, it's estimated that eventually more than 140,000 in Hiroshima died. As John Hersey wrote in his epic feature article in the entire August 31, 1946, issue of the *New Yorker*: "After the terrible flash—which, Father Kleinsorge later realized, reminded him of something he had read as a boy about a large meteor colliding with the earth—he had time (since he was 1,400 yards from the center) for one thought: A bomb has fallen directly on us." Even though US President Harry Truman vowed that the Japanese could "expect a rain of ruin from the air, the like of which has never been seen on this earth," the bomb wasn't enough to end the war. Three days later, on August 9, the bomb known as Fat Man was dropped on Nagasaki, which killed between 60,000 and 80,000 people. After this epic carnage, Japan surrendered on August 15, 1945.

Giles G. McCoy talked about what happened next, recorded by the Veterans History Project, after his group was rescued:

> Then they put us on the hospital ship *Tranquility*. They had to cart all of us out in stretchers, nobody hardly could walk. They put us on this real nice hospital ship and took us all back to Guam. They put us in Base 18 Hospital on Guam. That was great; we all knew that we had a good shot at living now. Most of us were recuperating. I remember one day the Captain came down and I was up. I had a young Marine right next to me and my bunk and he had lost his leg. He kept forgetting; he would get out of bed to go to the toilet and land on his stub and bust it open and start bleeding. I would jump out and we would put towels around him and pack him with towels to stop the bleeding and get him back into the operating room. He did this twice and finally they said they weren't going to operate on him any more; they weren't going to sew him up. He was going to have to behave. We pulled his bunk up real close to mine. I said, "Damn you, when

you have to go to potty, you holler and I will go with you." So we did; we would all take turns taking him back and getting him all fixed up so that he could urinate and go to the potty without busting his leg open.

So I was up doing this one morning when Captain McVay came in and he came up to me and said, "McCoy, I see you are up and around. Are you feeling alright?" I said, "Well, as good as I can, sir. What do you need?" He said, "You know what I need? I need a jeep and I need a driver. Would you be my orderly?" I said, "Sir, I would be happy to do that for you, anything to get me out of here." He said, "Alright, I've got a driver and a jeep out there now, have them take you down there [to the motor pool] and check out a jeep for me; just put it in my name." I said, "Alright, sir." So that is what I did and I drove him back and forth to CINCPAC which was the mountain where Admiral Nimitz and all of them had their headquarters on Guam. CINCPAC means Commander In Chief, Pacific.

McCoy said that Captain McVay later told him:

"You know what, McCoy? I think they are going to try and hook me for the loss of the ship." I couldn't believe it. That's all he ever told me, he never did confide in me. I said, "I don't know how they can do that sir. Gosh, you are a survivor just like us." I said, "That was an act of war. Hell, we know we got sunk by torpedoes and but it was an act of war." Years after they did court-martial him and then years after (back in 1990 after some of the books got out about the *Indianapolis*) the government sent my wife and me over to Pearl Harbor and I got a chance to meet Hashimoto [the captain of the Japanese submarine that sank the *Indianapolis*]. They brought him over from Japan to Pearl Harbor. They also brought him to the States to testify against Captain McVay. I know that whenever they did that, to me . . . I had my testimony; I was a witness at the trial of Captain McVay, so I was in Washington DC. When it was my turn to go into the court-martial area to testify they sat me right next to Hashimoto. I couldn't believe it! I just was so upset! I even said so, and Captain McVay, he liked that. I said, "How come you got . . . Who is this guy here? He is Japanese." They said, "Well, he is the one who sunk you." I said, "How can you do that? Is he testifying against my skipper?" I said, "For crying out loud, my skipper is a great man! Why did you bring him over for?" They told me that it was none of my business, it was part of the court-martial. I know that Captain McVay, he liked what I said. When I got through with my testimony I went back and I told the rest of the guys, I said, "You know, they got that Japanese bastard that sunk us. You are going to have to sit next to him when you testify." That upset everybody, too. Anyway, we went through it but they court-martialed him. We all felt that it was such a terrible thing to do.

Well, in 1960 I got the [survivor] reunion going and he [Captain McVay] came to our 1960 reunion (the first one) and I asked him for permission to try and get him exonerated. He refused me. He said, "No, I was the commanding officer and I will take my punishment." I said, "Well sir, it is unjust." He said, "That's alright, that's the way the Navy works." So he wouldn't give me permission. Then in 1964 (we had [reunions] every five years) we were getting ready for the '65 reunion and I called him and said, "Skipper, will you give me permission to try to get you exonerated?" He said, "I can't go to the reunion because my wife is dying, but I am going to give you permission but it isn't going to do any good. Don't work too hard on it because the Navy won't back off." I said, "But you will allow me to do it?" He said, "Yes, go ahead." So I started and didn't get anywhere until that young boy up in Pensacola, Florida, little Hunter Scott got started. God love him! He called me and I told him how far I had gone and gave him a bunch of names of guys that I thought would talk to him because some of the guys wouldn't talk about it. Some of them even busted me for getting the reunions going. They thought that was wrong, bringing it [the memories] back.

Hunter Scott was a twelve-year-old student in Pensacola, Florida, and as part of a school project for the National History Day program, he interviewed more than one hundred survivors of the *Indianapolis*. Scott testified to US Congress, which shone a national spotlight on McVay's plight.

Captain McVay had committed suicide in 1968 in Litchfield, Connecticut, on the steps of his house. The words "zigzag" might have been too burdensome—McVay was haunted, traumatized by the Secretary of the Navy James Forrestal's words "hazarding his ship by falling to zigzag" every time he watched a housefly buzz in its zigzag flight around the stovetop in the kitchen. McCoy said: "Captain McVay was a great skipper, and we all had tremendous respect for him. We really knew he was top drawer, as far as any of us were concerned. He wanted a good ship, he wanted a well-disciplined ship. I had tremendous respect for him. We got him exonerated and I was just overjoyed with the fact that we got him exonerated. I just wish that he was alive to understand that we didn't give up on him, that we stayed with him. That is what I tried to get the Navy and all to understand, that combat people just don't forsake other combat people. We stay there and we fight for honor. That is what was violated with Captain McVay; his honor was violated. It was up to us to fight to get his honor back. That is what we did and I am proud that we did." President Bill Clinton signed the resolution in 2000, clearing McVay's record.

McCoy described giving an address to schoolchildren about his World War II experiences:

> World War II veterans saved the world. If it wasn't
> for the men in Europe beating Germany and for
> us in the Pacific beating the [Japanese] our world

would not be free. Our world would be dominated and you wouldn't have the freedoms that you all have right now. Remember this and when you go a little further think back to whenever our fore-fathers started this country, when they made the Declaration of Independence, think of what that great document meant and think of the men back then that gave their lives to see that it was held up and that people didn't destroy those ideas. I always tell the kids, "I am proud that I had a chance to serve my country that needed me because it is a privilege to live in this country, so don't ever forget that." I don't want any of us old people to forget it. It is a privilege to be here and to live in this country.

Sitting in Umenomiyn Shrine in Kyoto in 2000, Mochitsura Hashimoto imagined the thoughts Captain McVay had when he was called to trial: *American Navy had to blame someone, and it is what we military officers accept, our lives would go down with memory of ship, forever we would be part of sunken wreck at bottom of sea.* Hashimoto felt no remorse for sending the torpedoes—he did what he was trained to do and what he pledged he would do for the Emperor. But still, he felt a sorrow for the plight of McVay, though he also felt that a man of his stature and fortitude should persevere, regardless of the outcome. Sadly, he had heard the news of McVay's suicide, and he understood the despondency and McVay's inability to for-give himself for the loss of many hundreds of lives, but he did not understand McVay's ultimate decision. One more death and more suffering of family was not the choice.

He believed McVay was never wrong in his handling of the *Indianapolis* and the aftermath—*he did not know a submarine lurked near, as war allowed*—and that in the military you simply have no choice and all you can do is be honest in your efforts to protect the men in your command. Hashimoto had done that—though he was never tested like McVay.

Hashimoto could feel the energy around him, the beauty of nature, and he sat placidly and filled with reverence in the shrine. *War and the brutality to which we commit does not have to change our true character*, he thought, reflecting on his own service. He believed, as he wrote in his book *Sunk*: "[T]he martial spirits of its sailors are still with us on the far-flung oceans . . . [W]e remember the multitude of resentful sleeping warriors; in our ears we hear him whisper of the 'voice from the bottom of the sea.'"

He knew, as he testified during McVay's court-martial trial in 1945, that no navigational tricks would have saved the *Indianapolis* from the *I-58*'s torpedoes—from Hashimoto's torpedoes. McVay might have zigged and zagged, but the result would have been the same, the final naval conquest for the Japanese nation and a horrific end for hundreds of US soldiers and sailors. Zigging and zagging would not have saved the *Indianapolis* and her crew, Hashimoto testified.

Later, in 1990, Hashimoto was invited to a reunion of *Indianapolis* survivors in Pearl Harbor. He told Dr. Giles McCoy, through a translator, "I came here to pray with you for your shipmates whose death I caused." Dr. McCoy responded: "I forgive you."

A report in the *New York Times* on July 14, 2001, stated: "On Nov. 24, 1999, a year before his death, Mr. Hashimoto wrote to Senator [John] W. Warner. 'Our peoples have forgiven each other for that terrible war,' he said. 'Perhaps it is time your peoples forgave Captain McVay for the humiliation of his unjust conviction.'"

Back in his present moment in the Kyoto shrine, Hashimoto called to mind the prayer he heard repeated during his days in Washington DC in the United States at Captain McVay's trial—the prayer that sustained so many and was cited over and again by Dr. Lewis Hayes, the ship's doctor and a survivor of the sinking who saved countless others; it was offered in the testimony at the trail of Captain McVay. This was the prayer that kept Hayes alive, he said, but also now kept him remembering the horror of those days in the water, the boys killed by sharks and those who went mad and could not sustain during the days and night in the water and on a raft, those who let go and were delivered away to their final rest—the Lord's Prayer.

Months later, in October 2000, Captain Charles Butler McVay III was posthumously exonerated. President Bill Clinton signed the Senate Resolution in October, which cleared McVay's Naval record. The Senate Resolution read: "Whereas, In a highly controversial proceeding, Captain Charles McVay III, the captain of the USS *Indianapolis* and one of the survivors, was court-martialed and convicted of 'hazarding his ship by failing to zig-zag', making him the only captain to be court-martialed for losing a ship in combat; Whereas, In October 2000, President Clinton signed legislation exonerating Captain

McVay and the Navy conceded that Captain McVay was innocent of any wrongdoing; whereas, the sinking of the USS *Indianapolis* remains the worst U.S. Naval disaster in history and the worst loss of life from shark attack in naval history."

Mochitsura Hashimoto, the commander of *I-58* who played no small part in Congress's reconsideration of McVay's court-martial, died in Kyoto on October 25, 2000, five days before the United States' historic exoneration of McVay for the plight of the *Indianapolis* was made official; the Shinto priest was ninety-one.

# THE WRECK OF HMS
# *BIRKENHEAD*

*The following are historical accounts of the
sinking of the HMS* Birkenhead *in 1852.*
From *The Times* (London), April 8, 1852

## LOSS of the TROOPSHIP *BIRKENHEAD*

The following dispatches and enclosures were yesterday received by the Board of Admiralty from Commodore
Wyvill, containing the painfully-interesting details connected with the loss of this ill-fated steamer. It appears that
the total number of lives which have been lost on this sad
occasion amounts to 438.

Castor, Simon's Bay, March 3. "1. Sir, It is with much
pain I have to report, for the information of my Lord
Commissioners of the Admiralty, the disastrous wreck of
Her Majesty's steam troopship Birkenhead, on the morning
of the 26th of February, at 2 o'clock, on the reef of rocks off
Point Danger, about 50 miles from this anchorage, by which
catastrophe I lament to state that the lives of 438 officers

(naval and military), women, soldiers, and boys, have been lost, out of 680 who were on board the ship at the time. The circumstances, as far as I can collect, are as follows:

"2. The Birkenhead reached Simon's Bay, from Cork, on the 23rd, in 47 days. She was immediately prepared for sea, received coals to 350 tons, some provisions, and the officers' horses, disembarked the women and children (except those taking a passage to Algoa Bay), and was reported ready on the 25th. That afternoon Mr. Salmond received the Government dispatches for his Excellency Sir Harry Smith, my orders to proceed to Algoa Bay and Buffalo Mouth to land the draughts of the different regiments, and stemmed on his passage at 6 o'clock in the evening, which was fine and calm, with smooth water.

"3. At half-past 2 o'clock on the afternoon of the 27th of February Mr. Culhane, assistant-surgeon of the Birkenhead, arrived at Simon's Town by land, to report to me the loss of his ship near Point Danger—that two boats, with, as he stated, the only survivors, were cruising about at a distance from the land."

From the *Albany Evening Journal*
(Albany, New York), April 20, 1852

Another terrible disaster has happened at sea. At 2 o'clock in the morning of the 26th of February her majesty's steamer, the *Birkenhead*, was wrecked between two and three miles from the shore of Southern Africa. The exact spot at which the calamity happened was Point Danger. Off this point she struck upon a reef of sunken rocks. The ship was steaming eight and a half knots at the time. The water was smooth, and the sky serene, but the speed at which the

vessel was passing through the water proved her destruction. The rock penetrated through her bottom just aft of the
foremast, and in twenty minutes time there were a few floating spars and a few miserable creatures clinging to them,
and this was all that remained of the *Birkenhead*. Of 633
persons who had left Simon's Bay in the gallant ship but a
few hours before, only 184 remain to tell the tale. No less
than 454 Englishmen have come to so lamentable an end.

There is not mystery about the calamity. We are left, as
in the case of the *Amazon*, to conjecture the origin of the
disaster. Just what happened to the *Orion* off the Scottish
coast, or to the *Great Liverpool* off Finisterre, has happened now. Captain Salmond, the officer in command,
anxious to shorten the run to Algoa Bay as much as was
possible, and more than prudent, hugged the shore too
closely. Four hundred and fifty-four persons have lost their
lives in consequence of his temerity. As soon as the vessel
struck upon the rocks the rush of water was so great that
the men on the lower troop-deck were drowned in their
hammocks. Theirs was the happier fate; at least they were
spared the terrible agony of the next twenty minutes. At
least the manner of death was less painful than with others, who were first crushed beneath the falling spars and
funnel and then swept away to be devoured by the sharks,
who were prowling around the wreck. From the moment
the ship struck, all appears to have been done that human
courage or coolness could effect. The soldiers were mus-
tered on the afterdeck. The instinct of discipline was
stronger than the instinct of life. The men fell into place
as coolly as on the parade ground. They were told-off into
reliefs, and sent—some to the chain-pumps, some to the

paddle-box boats. Captain Wright, of the ninety-first regiment, who survives to relate the dreadful scene he tells us:

"Every man did as he was directed, and there was not a cry or a murmur among them until the vessel mad her final plunge. I could not name any individual officer who did more than another. All received their order, and had them carried out as if the men were embarking instead of going to the bottom; there was only this difference—that I never saw any embarkation conducted with so little noise or confusion."

Poor fellows! Had they died in battle-field and in their country's cause, their fate would have excited less poignant regret; but there is something inexpressibly touching in the quiet unflinching resolution of so many brace hearts struggling manfully to the last against an inevitable disaster. It is gratifying, also, to find that the women and children were all saved. They had been quietly collected under the poop awning, and were as quietly got over the ship's side, and passed into the cutter. The boat stood off about 150 yards from the ill-starred *Birkenhead*, and all were saved. There is not the name of a single woman or child upon the list of persons who perished. The other boats, as is usual in such cases, were not forthcoming in the hour of need. One gig and two cutters were all that could be rendered available. In one account we find that when the men were ordered to get the paddle box boats out, the pin of the davits was rusted in, and could not be got out. Captain Wright, on the other hand, tells us that when the funnel went over the side it carried away the starboard paddle-box and the boat, and the other paddleboat capsized as it was being lowered. Of the 184 persons who were saved, 116 made their escape in the three boats which succeeded in getting clear of the wreck . . .

Never was destruction more sudden or more complete. Within fifteen minutes after the vessel struck the bow broke short off. Five minutes more elapsed, and the hull of the vessel went in two, crossways, just abaft the engine room. The stern part of the vessel immediately surged, filled and went down. The only hope of the survivors lay in the main topmast and main topsail yard, which still showed above water. There were some fragments of the forecastle deck still floating about; there were a few spars, and the driftwood. About forty-five people clung to the yard, and after remaining there until 2 o'clock the following afternoon, were picked off by the *Lioness*, a schooner which was providentially at hand.

Captain Wright asserts, that of the 200 persons, more or less, who were clinging to the drift-wood when he got away, nearly every man might have been saved had one of the ship's boats done her duty. Into this boat the assistant surgeon had got, with eight men. They immediately pulled away, and landed about fifteen miles from the vessel. The fact appears to have been, that the poor creatures who were clinging to the drift-wood had been carried by the swell in the direction of Point Danger. There they go entangled among the seaweed, which, at this point of the coast, is thick and of immense length. Captain Wright's opinion is, that had not the assistant surgeon carried off the boat, or even had the boat pulled back to the scene of the disaster, after landing the medical gentleman in safety, the majority of these persons might have been picked off the seaweed. It only remains for us to mention here that Captain Salmond, who appears to have done his duty after the vessel struck, had not survived the calamity. When last seen alive, he was swimming from the sternpost of the ship, which had just gone down, to a portion of the forecastle-deck, which was floating about twenty yards from the main body of the wreck; something struck him on the back of the head, and he never rose again.

From *The Times* (London), Thursday, April 8, 1952

## "REPORT OF THE MASTER OF THE SCHOONER *LIONESS*.

## "ENCLOSURE NO. 2, IN NO. 12 OF 1852, FROM COMMODORE WYVILL.

"Schooner *Lioness*, Simon's Bay. Feb 27.

"Sir.—I beg to report to you that when the *Lioness* was off Walker's Bay I observed a boat in the shore pulling towards me on the morning of the 26th inst., at about 10 o'clock, which I picked up, and found her to contain 37 survivors from the wreck *Birkenhead*. On hearing that there were two other boats I proceeded in search of them, and three-quarters of an hour afterwards I succeeded in picking up the other cutter with the women and children; but, after cruising about for the third boat, I made for the wreck, which I reached about 2 o'clock in the afternoon of the same day, and sent the cutters away to pick up the men handing to the spars, by which we rescued 35 soldiers and sailors in a nearly naked state. The wreck had disappeared all but a piece of the main topmast and top sailyard, to which the men were clinging. Nothing else could be saved."

From *Albany Evening Journal* (Albany, New York),
Wednesday, April 28, 1852

**The Loss of the *Birkenhead*:**

The following is an extract of a letter from Lieutenant Girardot, 48d Light Infantry:—

"Simon's Bay, March 1.

"My Dear Father:—I wrote on letter to say I was safe, but for fear that it should not reach you, I will send this to say I am quite well. I remained on the wreck until she went down the suction took me down some way, and a man got hold of my leg, but I managed to kick him off and come up, and struck off some pieces of wood that were on the water, and started for land, which was about two miles off. I was in the water about five hours, as the shore was so rocky, and the surf ran so high, that a great many were lost trying to land.

Nearly all those that took to the water without their clothes on were taken by sharks; hundreds of them were all around me, but as I was dressed, (having on a flannel shirt and trousers) they preferred the others. I was not in the least hurt, and am happy to say kept my head clear; most of the officers lost their lives from losing their presence of mind, and trying to take money with them, and from not throwing off their coats. There was no time to get the paddle-box boats down, and a great many more might have been saved, but the boats that were got down deserted us and went off. From the time she struck to when she went down was 20 minutes.

When I landed I found an officer of the 12th Lancers, who had swam off with a life-preserver, and 14 men, who had got on with bits of wood like myself. We walked up the country 11 miles, to a farm belonging to Captain Smales, formerly of the 7th Dragoon Guards, who was very kind to use, and all the men that were got on shore came up to him. I hope the Government will make up our loss to us, as I have saved nothing. Melford, of the 6th, the ensign I spoke of as having a wife on board with him, went down; she, poor thing, was here when the ship sailed for Buffalo Mouth; I have just been to see her, and she looks more dead than alive, left all alone at this distance from her home, but we shall do all we can to be of service to her. There is a report that many have been killed in the Amatola Mountains, and our poor doctor was killed some little time back. God grant that we may be all spared to meet again. Ever your affectionate son,

—FRANK GIRARDOT

# THE SINKING
# OF *LA SEYNE*

*In the early morning of November 13, 1909, in the dark-
ness of a new moon, the 2,379-ton* steamer La Seyne
*collided with the steamer* Onda *and went down quickly—the
news of the disaster was sent around the world by wire service or by news files on board other ships coming to port, and
some reports identified the steamer as* Laaseyne, Lazyne,
*or* Laseyne *and it had previously been in service named*
Etoile du Chili. *Passengers were killed from the initial
explosion of the collision, drowned in the Rhio Straight
about thirty miles from Singapore, and were devoured by the
prowling sharks. Eventually, sixty-one survivors made it to
the* Onda.

From the *Des Moines Register*, Monday,
November 15, 1909

100 DROWN IN COLLISION
TWO BIG MAIL STEAMERS COLLIDE AND SINK.
Sixty-One Rescued From Jaws of Sharks, Which Cavorted
About in Shoals.

SINGAPORE, Nov 14—The mail steamer *La Seyne* of the Messageries Maritimes service, running between Java and Singapore, and on her way to this port, was in collision early this morning with the steamer *Onda* of the British-India line, and sank within two minutes. Seven European passengers, including Baron and Baroness Benicza, the capain of *La Seyne*, five European officers and eight-eight others, comprising native passengers, and members of the crew, were drowned. The rescue of sixty-one persons, practically from the jaws of shoals of sharks, formed a thrilling incident of the wreck.

The accident occurred about 4 o'clock in the morning in a thick haze. The vessels were steaming at good speed and *La Seyne* was cut almost in half. There was no time for panic; not for any attempt on the part of the officers of the foundering steamer to get out of the boats. The majority of those on board were caught in their berths and carried down with the vessel.

The force of the collision brought the *Onda* to almost a dead stop and her engines were at once slowed and boats lowered. The rescue work proved thrilling, for not only were the rescuing parties impeded by the darkness, but shoals of sharks were already attacking those clinging to pieces of wreckage in the water. Sixty-one persons from the ill-fated steamer were finally dragged into the boats and brought by the *Onda* to this port. Many of them had been bitten by sharks and several are severely injured.

Headline from the *Washington Post*,
Monday, November 15, 1909

## SHIP SUNK; 101 DROWN

Only 61 Are Saved From *La Seyne* After Collision.

. . . Survivors Bitten by Sharks.

The rescue work proved thrilling, for not only were the rescuing parties impeded by the dark, but whole shoals of sharks were already attacking those clinging to pieces of wreckage in the water . . .

From *Zeehan and Dundas Herald*,
Tasmania, December 9, 1909

The files received by the steamer *Guthrie*, which arrived from Singapore yesterday, also give an account of a disastrous collision between the steamers *Laseyne* and *Onda* on 14th November. The collision occurred in the Straits of Rhio, about 28 miles from Singapore. When the two ships collided the French steamer when to the bottom in less than five minutes, and out of 154 people on board 93 perished. She went down by the head, and the survivors were left in the water with the only clothes they happened to be wearing. Commander Captain Conaflhoe lost his life. There was no time for orders to be given, or for the boats to be lowered. All of those rescued were picked up by three boats which were promptly put out by the *Onda*. Sharks abound in these waters, and it is believed that some of those people who got clear of the ship were attacked by sharks. Just as a Malay seaman was being dragged into one of the *Onda*'s boats a shark seized his foot, but a second engineer beat the brute off with a boat hook.

From the *Advertiser* (Adalaide, South Australia),
Thursday, December 9, 1909

Reprinted from National Library of Australia.

# THE WRECK OF THE LA SEYNE.

## COLLISION NEAR SINGAPORE.
## 93 LIVES LOST.

The "States Bridget" publishes the following account of the disastrous collision between the *La Seyne* and the *Onda* on November 14:

We regret to have to announce that in the early hours of Sunday morning there occurred in the Straits of Rhio, at a spot approximately 28 miles from Singapore, a disastrous collision involving the loss to the Messageries Maritimes steamer *La Seyne*, which keeps up a regular fortnightly connection between the outward and homeward French mail steamers calling here and at Batavia. The *La Seyne* was run into by or ran into the British India Steam Navigation Company's steamer *Onda* and sank almost instantaneously, carrying to their death 93 persons out of a total of about 154 souls aboard. The scene of the disaster the Rino Straits, it should be explained, form the main shipping highway for ships sailing between this port and Java. They are well lighted, but navigation is difficult, owing to the strong acts of the current, and great care has always to be exercised in negotiating the channel, especially when other ships are in the neighborhood. The *La Seyne* was travelling northward to Singapore. The *Onda* had sailed from this port on Saturday night, and was bound for Tegal, in Java. The two vessels approached each other at a spot where the strait is about two miles wide, near the lighthouse on Pulau San.

What actually caused the disaster is a matter which will be investigated at a court of enquiry to be held later.

It is natural that the officers on both sides should be ret-
icent on this point. They will make their statements at
the proper time. When the two ships collided the French
steamer went to the bottom in less than five minutes and
she now reste [*sic*] on the Pulau Sau side of the channel,
with some 25 ft of her mainmast as the only visible sign of
where she lies.

She went down by the head and the survivors were
left in the water with only the clothes they happened to
be wearing at the moment. The commander of the *La
Seyne* Captain Conaflhoe, lost his life. So suddenly did
the catastrophe happen that there was apparently no time
for orders to be given on the French ship or for boats to
be lowered. The vessel went down like a stone and it is
quite evident that the majority of those lost must have
been drowned like rats in a trap. A good many of the few
who managed to scramble on deck must have been imprisoned under the ship's awnings, and it is also clear from
what follows that of those who got clear of the ship a good
many were the victims of the sharks in which those waters
abound. All of those rescued were picked up by three
boats, which were promptly put out by the *Onda*, but it is
unhappily clear that if any remained alive who were not
picked up at the moment, they must have met their death
later from the sharks or from drowning.

### A Sailor's Account.

Among the European passengers on the *La Seyne* bound
for this port were six sailors who had been paid out from
their vessel, the *Daylight*, at Batavia. Of these D. Driscoll

and G. Craig had not been heard of and there is no doubt that they have been drowned. The other four (P. Bolton, H. Muller, C. Glendinning, and another) are now at the Sailors' Home. They have lost everything they possessed.

Mr. Glendinning on Monday told a representative of this paper what he heard and saw, and his account being that of a seafaring-man is likely to be more substantially correct than that of a passenger. Mr. Glendinning says that just after 4 o'clock on Sunday morning the *La Seyne* was steaming close to the Pulau Sau light. There had been a heavy thunderstorm the previous afternoon with plenty of rain, and the weather was hazy, though it was not actually raining. He and his mates had retired, but Mr. Glendinning hearing the ship's whistle give one blast proceeded to make his way upon deck, the six men being quartered a little forward of and below the bridge. He had not reached the deck when the *La Seyne* gave a couple more blasts, and almost simultaneously the two ships went into each other with a crash. The appalling suddenness with which the French ship floundered may be gathered from the fact that Mr. Glendinning says that he at once noticed she was sinking by the head. He immediately rushed to his mates and called them up, and on reaching deck again, shouted to the *Onda* people to throw out some lines, but added Mr. Glendinning, "there was no time for lines, she just sank in about 6 minutes from the time she struck. When the cold water got to her boilers they exploded and burst out her sides and she went down like a stone. In another moment we were all in the water. It appeared to be still water where we went in, but a couple of hundred yards or so off we went into a rip and began to travel away. There were all kinds of wreckage in the water round me, including rats, one of which jumped on my shoulder. There was a good deal of shouting, but it did not last long. It was no use swimming against the tide; it was too strong. I kept in heading for the other steamer, and after I had been in the water some time I was picked up by the boat in charge of the second engineer of the *Onda*."

Mr. Glendinning confirms the statement that a good many of the people who were thrown into the water must have been pulled down by sharks. There were many of them about. Just as a Malay seaman was being dragged into the boat which rescued Mr. Glendinning, a shark seized the man's foot, but the second engineer beat the brute off with a boathook, and the Malay was saved, but he was so badly bitten that he was obliged to go into the hospital. Mr. Glendinning believes it was the sharks that got most of those who got free of the ship, but were not picked up. There were many dead bodies floating about afterwards, but after the *Onda* had done all the work it was possible for her to do, the survivors were brought on to Singapore, where they arrived about mid-day on Sunday.

Mr. Glendinning also bears out the theory that a good many may have been drowned under the awnings. He says that the *La Seyne* was fairly on the Pulau Sau side of the channel and that there was plenty of water. He was on the forecastle head, just before the disaster, and noticed when he went below that the weather was hazy, but it cleared up soon after the collision. He is certain that a good many of the natives lost their lives through paying too much attention to saving their belongings, but one Chinaman got his box into the water and clung on to it, and fortunately saved both his life and his box.

From the *Singleton Argus*, New South Wales,
Australia, January 1, 1910

The mail steamer Empire arrived at Sydney on Wednesday brings the latest files containing full particulars of the terrible disaster which occurred to the Messageries Maritimes steamer *La Seyne* in the Rhio Straits, brief particulars of which were received here by cable on November 15th last.

It appears that the steamer *La Seyne*, commanded by Captain Conaihoe, and one of the best equipped of the company's fleet, was two days out from Batavia, bound for Singapore. When near Pulo Sasu, and some 30 miles from Singapore, another steamer, the *Onda*, of the British India fleet, crashed into her during the darkness. The *La Seyne* was so badly damaged that she settle down immediately, and the passengers and crew, numbering over 150 souls, suddenly found themselves in the turbulent waters of the Rhio Straights, which are infested with man eating sharks. Had it been a matter only of shipwreck many might have been saved, but, horrible scenes were enacted, the drowning people being seized by the sharks and dragged under, amidst heartrending scenes. Meanwhile, the *Onda*, a vessel of 3,409 tons, which was bound in ballast for Tegal, in Java, stood by, and great promptitude was shown in getting away three boats to the rescue; the French steamer having been unable to lower away any at all. The crew of the *Onda* worked with superhuman energy, and were successful in picking up seven of the passengers among whom were four British paid off seamen who were being sent to Singapore second class, 12 of the officers and European crew, 14 native passengers, and 28 native crew, or a total of 61 in all. The escape of steam from the ship's boilers, which seemed to open the boat up, scalded to death many of those on board. The survivors describe it as a frightful scene, but so terribly soon was it all over that there must have been many who went to their death without having time even to attempt to save themselves.

A passenger who was saved states that the sharks were very ferocious, attacking the women and children as well as the men. He saw a Chinaman fighting one of the huge monsters. After beating him off with a boat hook the Chinaman tried to get into the *Onda*'s boat, but the shark returned to the attack and bit off the man's foot. The captain, surgeon, and purser, as well as the chief cook of *La Seyne*, were amongst those drowned or eaten alive.

From the *Sun*, July 13, 1913

"The first of the passengers had scarcely touched the water before a shoal of sharks was circling the scene and dragging down scores of men and women who never came up again. These facts were sworn to by dozens of eyewitnesses to the spectacle."

# TERROR AT THE
# NEW JERSEY SHORE

We *don't mean modern terror or anxiety resulting from weekend road traffic around New Jersey tunnels or shopping malls . . . Back in 1916, nobody imagined that they would soon encounter the same kind of drama, suspense, and terror that decades later became a hit movie with Steven Spielberg's* Jaws. *But this terrifying week happened along the Garden State coast, in every horrific detail reported in these period stories.*

From the *New York Times*, July 13, 1916

NEW JERSEY, USA, 1916. Captain Thomas Cottrell, a retired sailor, caught a glimpse of a dark gray shape swimming rapidly in the shallow waters of Matawan Creek this morning [12 July] as he crossed the trolley drawbridge a few hundred yards from town. So impressed was he, when he recalled the two swimmers killed by sharks on the New Jersey coast within two weeks, that he hurried back to town and spread the warning among the 2,000 residents that a shark had entered Matawan Creek. Everywhere the

Captain was laughed at. How could a shark get ten miles away from the ocean, swim through Raritan Bay, and enter the shallow creek with only seventeen feet of water at its deepest spot and nowhere more than thirty-five feet wide? So the townsfolk asked one another, and grown-ups and children flocked to the creek as usual for their daily dip. But Captain Cottrell was right, and tonight the people are dynamiting the creek, hoping to bring to the surface the body of a small boy the shark dragged down. Elsewhere, in the Long Branch Memorial Hospital lies the body of a youth so terribly torn by the shark that he died of loss of blood, and in St Peter's Hospital in New Brunswick doctors are working late tonight to save the left leg of another lad whom the shark nipped as the big fish fled down the creek toward Raritan Bay.

The dynamiters hoped, when they brought their explosives to the creek, that, beside the body, they might bring up the shark where men, waiting with weapons, could kill it. Others hastened to the mouth of the creek where it empties into the bay a mile and a half from town and spread heavy wire netting.

The people of Matawan had been horrified by the tales of sharks which came to them from Spring Lake, Beach Haven, Asbury Park and the other coast resorts. They had been sympathetically affected by the reports of the death of Charles E. Vansant and Charles Bruder. But those places were far away and the tragedies had not touched them closely.

Tonight the whole town is stirred by a personal feeling, a feeling which makes men and women regard the fish as they might a human being who had taken the lives of a boy and a youth and badly, perhaps mortally, injured another youngster. The one purpose in which everybody shares is to get the shark, kill it and to see its body drawn up on the shore, where all may look and be assured it will destroy no more.

The death of the boy and youth, and the injury to the other youngster were due to the refusal of almost every one to believe that sharks could ever enter the shoal waters where clam-diggers work at low tide. As long ago as Sunday, Frank Slater saw the shark and told it everywhere. He stopped repeating the tale when everyone laughed him to scorn.

Then today came Captain Cottrell's warning, and with that Lester Stilwell, twelve years old, might have been the only victim had it not been for the unfortunate coincidence that the boy suffered from fits. It was supposed that an attack in the water had caused him to sink, and rescuers, with no notion that a shark had dragged him down, entered the water fearlessly.

It was while trying to bring young Stilwell's body ashore that Stanley G. Fisher, son of Captain W. H. Fisher, retired Commodore of the Savannah Line fleet, lost his life. The third victim, Joseph Dunn, twelve years old, was caught as he tried to leave the water, the alarm caused by Fisher's death at last having convinced the town that a shark really was in the creek.

Stilwell was the first to die. With several other boys, he had gone swimming off a disused steamboat pier at the edge of the town. He was a strong swimmer and so swam further out than his companions.

So it was that none could follow him, but several boys, instead, raced through the town calling that Stilwell had had a fit in the water and had gone down. They said the boy rose once after his first disappearance. He was screaming and yelling, and waving his arms wildly. His body was swirling round and round in the water. Fisher was one of the first to hear and immediately started for the creek.

"Remember what Captain Cottrell said!" exclaimed Miss May Anderson, a teacher in the local school, as Fisher passed her. "It may have been a shark."

"A shark here!" exclaimed Fisher incredulously. "I don't care anyway. I'm going after that boy."

He hurried to the shore and donned bathing tights. By the time he was attired many others had reached the spot, among them Stilwell's parents. Fisher dived into the creek and swam to midstream, where he dived once or twice in search of Stilwell's body. At last he came up and cried to the throng ashore: "I've got it!"

He was nearer the opposite shore and struck out in that direction, while Arthur Smith and Joseph Deulew put out in a motor boat to bring him back. Fisher was almost on the shore and, touching bottom, had risen to his feet, when the onlookers heard him utter a cry and throw up his arms. Stilwell's body slipped back into the stream and, with another cry, Fisher was dragged after it.

"The shark! The shark!" cried the crowd ashore, and other men sprang into other motor boats and started for the spot where Fisher had disappeared. Smith and Deulew were in the lead, but, before they overtook him, Fisher had risen and dragged himself to the bank, where he collapsed.

Those who reached him found the young man's right leg stripped of flesh from above the hip at the waist line to a point below the knee. It was as though the limb had been raked with heavy, dull knives. He was senseless from shock and pain, but was resuscitated by Dr G. L. Reynolds after Recorder Arthur Van Buskirk had made a tourniquet of rope and staunched the flow of blood from Fisher's frightful wound. Fisher said it was a shark that had grabbed him. He had felt the nip of its teeth on his leg, and had looked down and seen the fish clinging to him. Others ashore said they had seen the white belly of the shark as it turned when it seized Fisher.

Fisher said he wasn't in more than three or four feet of water when the fish grabbed him, and he had had no notion of sharks until that instant. If he had thought of them at all, he said, he had felt himself safe when he got his feet on the bottom.

Fisher was carried across the river and hurried in a motor car to the railroad station, where he was put aboard the 5:06 train for Long Branch. There he was transferred to the hospital, but died before he could be carried to the operating table.

At the creek, meantime, dynamite had been procured from the store of Asher P. Woolley and arrangements were being made to set it off when a motor boat raced up to the steamboat pier. At the wheel was J. R. Lefferts and in the craft lay young Dunn. With his brother William and several others, he had been swimming off the New Jersey Clay Company brickyards at Cliffwood, half a mile below the spot where Stilwell and Fisher were attacked.

News of the accident had just reached the boys and they had hurried from the water. Dunn was the last to leave and, as he drew himself up on the brick company's pier, with only his left leg trailing in the water, the shark struck at that. Its teeth shut over the leg above and below the knee and much of the flesh was torn away.

Apparently, however, the fish had struck this time in fright, for it loosed its grip on the boy at once, and his companions dragged him, yelling, up on to the pier. He was taken to the J. Fisher bag factory nearby, where Dr. H. J. Cooley of Keyport dressed his wound, and then he was carried in a motor car to St. Peter's Hospital in New Brunswick by E. H. Bomick. There it was said last night that the physicians hoped to save his leg if blood poisoning did not set in.

The youngster steadfastly refused to tell where he lived, for, he said, he did not want his mother to worry about him. From his relatives, however, it was learned that his home is at 124 East 128th Street, New York. He and his brother had been visiting an aunt in Cliffwood.

Fisher was the son of Commodore Watson H. Fisher, who for more than fifty years commanded boats of the Savannah Line up and down the coast. He retired from active service a few years ago. About ten days ago the father and mother went to Minneapolis to visit a daughter there, and they had intended to remain for another week, but, when word was sent this evening of the death of their son, they sent a message that they would leave for home immediately.

News of the tragedies here spread rapidly through neighboring towns, and from Morgan's Beach, a few miles away, came a report that two sharks had been killed there in the morning by lifeguards. One was said to be twelve feet long. Persons who saw the fish when it grabbed Fisher said they thought it was about nine feet long.

*The horrific news of this attack spread quickly. A local newspaper in Brooklyn, New York, dedicated an entire page to the subject of sharks.*

From the *BROOKLYN DAILY EAGLE*,
Friday, July 14, 1916

The recent activity of sharks along the Jersey coast and the fact that one was sighted off Mecox Beach here a few days ago by Philip Carter, a nephew of ex-Justice Hughes, has brought to mind that in years past the Atlantic coast, from Montauk to Shinnecock, was the chief fishing grounds.

In inquiring the probability of there being any man-eating sharks in these waters now a number of the oldest inhabitants of Bridgehampton say that they remember the time when they inhabited the water about here in large numbers.

Captain John Norris Hedges, who has lived in Bridgehampton for sixty-nine years and was the captain of the Mecox lifesaving crew for thirty-nine years, says that it was not uncommon about forty years ago to catch sharks of man-eating variety off Bridgehampton Beach; but owing to the fishing the last few years (which has driven the bunkers in small fish which is a favorite food of sharks) farther out to sea, the sharks have not come so far inshore of late until this season, which he attributes to the fact that there has been little fishing the past few years, and that the bunkers are coming closer in shore, and that the sharks are following them.

E. E. Halsey, another resident of Bridgehampton, who has passed a good part of seventy-nine years in this locality, corroborates Captain Hedges' belief, and say that it is very likely that man-eaters are again inhabiting these waters.

Charles Deckert was a member of the crew that landed the last shark in this vicinity, which was of the blue-nosed variety and was brought to shore in the winter of 1914. It measured nine feet.

An unusual number of porpoises have been sighted near shore here of late, and on Sunday and Monday two whale about forty feet long were seen basking off Sag Beach.

Most Sharks on Long Island
Belong to Basking Family

Patchogue, L. I., July 14—Baymen and deep sea fisher who have had experience in shark hunting have great respect for wounded sharks. Captain Frank Rourke, who once shot a shark through the fin, described its performance:

"It shot ahead as straight and as swift as a cannon ball, for a quarter of a mile, lashing the water in an intense fury. Suddenly, like a shell hitting a wall of rock it stopped, and then with increased speed shot back in its own tracks, narrowly missing the boat from which the shot had been fired. So great is the speed of a shark enraged that it would wreck the stoutest skiff that came in its path.

"Most of the sharks in and around the bay are of the basking shark family, but there are over a half dozen varieties known here. Among the dog fish, which are young sharks, monsters fifteen feet in length have been observed several times in the bay.

"There is but one adversary that is feared by the shark that is the dolphin or porpoise. A dolphin will give battle to a shark and slash it to pieces by turning somersaults in the water so rapidly that its finds cut like a buzz saw. A hungry porpoise is in the same class as a shark.

"Sharks mother their young in the same manner as a kangaroo, and baymen thought practical the plan to establish a guard at Fire Island Inlet and shoot the sharks as they come into the bay with their young and altogether make it such an unhealthy place for them that they will avoid it and food fish find a haven.

"The disappearance of moss bunkers is believed to have put the sharks in sore straights for food and caused their depredations."

Old Whalers Incredulous;
Man Eaters Only in Tropics

(Special to the *Eagle*.)

Sag Harbor, L. I. July 14—Old whalemen interviewed cannot recall any instance of sharks attacking men in waters other than the tropics. They are incredulous of stories of man-eating sharks doing mortal damage in small creeks or even along the beaches of New Jersey and Long Island. Gus De Castro Jr. and Stewart Gaffga, while bathing last summer at Cockles Harbor, Shelter Island, encountered sharks on the flats. They returned in their powerboat to Sag Harbor and armed themselves with whalers' harpoons and lances. They killed two big sharks . . .

Man-Eating Sharks? Nonsense!

## No Such Thing, Say Fishermen

Editor Arthur Knowlson of *Fishing Guide* Knows
All About Sharks' Habits—Has Seen Many, but No
"Man-Eaters" Around These Waters—Others Say
the Same.
They're Astonished at New Jersey Stories.

According to the men who do down to the sea in ships each day for the purpose of catching fish, there is no record of a shark attacking a human being in Long Island waters. Some of the fisherman go as far as to say that there is no record in existence of a shark ever attacking a human being in Northern waters, and point to the offer made by the late Herman Oelrich of Manhattan of $500 for anyone who would bring him absolute proof that a shark had attacked a living man at sea. This sum was never won, although Mr. Oelrich received many stories, with the necessary proof.

Arthur Knowlson, editor and publisher of the *Fishing Guide*, "Fishing Around New York," and similar publications, stated today that in all the years he has been fishing in Long Island waters, and that extends over a period of twenty-five years, he has never known of a shark attacking a human being.

"I can offer no explanation of what had happened off the New Jersey coast," said Mr. Knowlson. "Undoubtedly, the swimmers were attacked by some sort of a fish, and as there is no other fish in the seas that could do such damage, it must have been a shark that was responsible. But it is at such odds with the usual customs of the fish that it is hard to believe.

### Plenty of Sharks in New York Waters

"We have plenty of sharks in New York waters. There is the sand shark, the sucker shark and the hammer head. I have seen them upon many occasions, when I have been out in my canoe on Gravesend Bay. You will see the dark dorsal fin of the fish cutting through the water like a periscope. You can always tell a shark, as it goes steadily about its business. A porpoise tumbles through the water in big half circles, and so it could not be mistaken for a shark.

"I have had a shark come along, dive underneath my canoe and come up on the other side of the boat. These fish have been here for years. They run in years. Sometimes there are many of them in neighboring waters, and at other times they seldom, if ever, are seen. Last year was quite a season for sharks; in fact, there were so many of them around New York that one enterprising owner of a fishing boat advertised shark fishing, and only went out after the big fellows. This year there have been fewer reports from the Fishing Banks.

"The shark is a natural born scavenger. Although he lives off all forms of fish, he will eat it either dead or alive. The men who go after sharks generally fish with the head and the inside of some other fish as bait.

"The fishermen can always tell when the sharks are hungry. They will come alongside a boat, and when a weakfish or a blue or some other fish is hooked, he will turn on his side and bite the fish right off the hook. Many times I have had a shark clean my hook in this manner. Sharks have done very little of this stealing this summer. That makes me think that they have plenty of natural food. There, it is very hard to understand why they should attack human beings."

### No Record of a Shark in
### L. I. Waters Attacking Man.

Mr. Knowlson was asked if the sharks that are usually around Long Island waters would attack a man.

"No record of such an attack has ever been known," was the reply. "Still, if fish were hungry, I would not put it past him. But, judging by what happened at the fishing banks this year, there is no famine of shark food around here this year."

If the fish that attacked the swimmers off the New Jersey coast is not the usual type of shark common to these waters, do you think it is a man-eater from the South?" was the next question put to the fishing authority.

"Now we are drifting into the question of man-eating sharks," was the reply, "and there is also a grave doubt if there really exists such as fish. There is no record in either the United States navy or the English navy of a shark attacking a sailor. When you stop to think that both American and English sailors swim in every clime, it is a rather astonishing fact that neither navy has any record of a man being attacked by a shark.

"Havana harbor is supposed to be filled with sharks. Sill the Cubans go in swimming with impunity. When American warships visit Cuba the men go overboard in so-called shark-infested waters, and nothing seems to happen. Some people state that a shark will not attack a human being unless he is perfectly motionless. Other authorities say that the fish will not attack a man in the nude. Both of these statements seem to be wrong if the shark stories from the Jersey coast are true. Boys swimming in the nude were attacked yesterday."

"Have you any idea what would bring a so-called man-eating shark to northern waters?" "Lack of food in the South," was the reply.

"Do you think that the great loss of life caused by the sinking of so many boats abroad would cause a shark to come north after dead bodies?"

### Shark a Surface Feeder,
### Mr. Knowlson Points Out.

"No. The shark is a surface-feeder. He takes all his food within a few feet of the top of the water. This is proven by the fact that when a school of small fish want to get away from a shark they all dive down to the bottom. The body of a drowned man sinks to the bottom, and so could not furnish food for sharks."

"Are sharks terrible fighters?" was another question put to Mr. Knowlson.

"Yes and no," said the authority. "There are many stories told by local fishermen of porpoises chasing a shark up into Jamaica Bay creeks and killing the so-called tiger of the sea. I have had other fishermen tell me that they have seen hammerheads kill porpoises."

## Never Heard of Attack On Man in L. I. Waters

P. I. Evans of the Osborn House, at Sheepshead Bay, a great rendezvous for fishermen, said that he never knew of a shark attacking a human being in Long Island waters. "We have plenty of hammerheads and other species of shark around here, and it is no uncommon thing for a fishing boat to bring in a big fellow. As a rule, they are simply brought to share as a curiosity, and the next day the skipper takes the dead fish out to sea and throws him overboard. These sharks are generally caught off the Cholera Banks or the Klondike. I do not remember any of them being caught nearer shore. As far as a man-eating shark is concerned, I never saw one or heard of anybody being attacked."

Joe Gillies of the Great Kills Hotel, a famous Staten Island fishing resort, laughed at the idea of a man-eating shark. "We seldom see sand sharks around here," said Captain Joe, "and as for a man-eater, such a thing is out of the question. Still, it is mighty strange what happened across there in Jersey."

Charles Noehren, who owns a fishing station at Goose Creek, Jamaica Bay, said today that he had read the shark stories, but did not believe that the fish would bother him. "We are so far up the bay that we never are troubled by big fish. If one gets up I will take a pop at him."

From the *Sun* (New York, New York),
Sunday, July 16, 1916, Special Feature
Supplement: "Tigers of the Sea"

## All Theories of Scientists Upset by Invasion of Local Waters by Man Eaters

By John Walker Harrington

With lazy fins he swims the ocean ways, this creature of the tropics deeps. And sometimes a victim slays upon the barges of our northern seas.

So rare is his coming that there prevailed a belief that he never killed in Atlantic coast waters north of Hatteras. That there was good ground for this feeling of security is evidenced by the fact that scientists insist that the past fortnight has produced the first authentic cases in these latitudes of human beings having been actually destroyed by sharks. The fate of Charles Bruder, the Swiss bell-boy whose legs were bitten off while he was in bathing beyond the surf line at Spring Lake, and that of Charles E. Vansant, who was mangled while swimming at Beach Haven by a monster whose glistening fins were plainly seen, have been widely discussed as showing that the sea wolf lurks along the New Jersey shores.

One man eater might have attacked both men, for although the waters off the eastern seaboard teem with some fifteen varieties of sharks both great and small—big sharks with small teeth and little sharks which look like animated saws—sharks of the kind that made the killings reported are only chance visitors in this quarter of the ocean. Even in the equatorial regions where the man eating shark dwells the tragedies laid to him are few.

The appearance of a shark in Matawan creek last Wednesday, where it killed a boy, Lester Stillwell, and a man, Stanley Fisher, who went to his rescue, and mangled another boy, Joseph R. Dunn, was an even greater surprise than the two previous attacks on bathers. Old fishermen said it was unheard of for sharks to travel so far inland.

So seldom does danger from sharks impend that the late Hermann Oelrichs, one of the strongest and boldest swimmers the New Jersey coast has ever known, made an offer in *The Sun* of a reward of $500 for a well authenticated case of a man having been attacked by a shark in temperate waters. The reward, which was offered in 1899, was never claimed. It stirred up much comment, however, and a general overhauling of natural history archives.

The nearest to a circumstantial account of a shark's seeking the life a man was that which was reported from Greenport, Long Island. A sailor who was swimming ashore from a ship near Horton Point was bitten on the hip by a supposed shark, and with great difficulty was rescued by fishermen. He suffered much loss of blood, and was revived when taken ashore.

The man was carried in a wagon to Greenport and treated by a surgeon, who succeeded in saving him after several months of weakness and exhaustion. It is unfortunate, however, that neither the name of the man nor that of the vessel is at hand; in fact, there is no official record by which this account can be substantiated.

The real man eater, the *Carcharodon carcharias*, will when the chance offers attack human beings. He ranges the seas seeking his prey, just as the lion roams the jungle or the wolf prowls through the woods. He makes his lair in the great Gulf Stream. At this season of the year the Gulf Stream beats to the northern Atlantic a host of the finny denizens of the southern waters, and with them come the sharks.

The great white shark, or man eater, is a cold blooded fish enough, but he seldom ventures into chilly water. When the Gulf Stream boils under the copper skies the sea wolf starts in his indolent zigzag course for the higher latitudes, keeping at first well within the limits of the tropic current. He and his kind are shown graphically in Winslow Homer's well known painting "The Gulf Stream" in the Metropolitan Museum of Art in this city. On the deck of a dismasted hulk is a negro, and in the water alongside sharks swim about with their mouths upturned waiting for his exhausted or dead body to slip into the water.

Sharks leave the Gulf Stream when they reach the northern seas if the water to the west of them is warm enough or they are drawn by the prospect of good food. The smaller varieties will enter the surf and even disport in the harbors, but the larger ones, being essentially pelagic creatures, keep in the deep water as a rule.

The great white shark grows to be from thirty-five to forty feet long. He is the strongest and the boldest of his genus. According to Linnaeus, he is that leviathan of Holy Writ who swallowed Jonah, and in the days of Nineveh he might have attained a far greater length than the specimens which have been observed in the seas of the New World.

The shark has changed little in appearance since the beginning of his history. The great white shark retains all the characteristics of the prehistoric variety from which he descended. He has the same reach of jaw and the deep set, triangular teeth of his brothers of the lost eons. There stands in the Museum of Natural History a remarkable head of a prehistoric shark in the jaws of which a man can easily sit and have room to spare.

There is much in common between this monster of the tertiary period and the ranger from the Gulf Stream, especially in the way in which both go after their prey. The ancient type was fully seventy feet in length and was undoubtedly the largest fish that ever swam the ancient seas.

The great white shark when young is white underneath and black and gray on the back. As he grows older he acquires the peculiar, sickly, grayish white which gives him his name. An evil spirit of the deep he seems, for there is something about him that fills man with that instinctive dread felt for the snake.

The body of the fish, shiny, eel-like, sinuous and marked by heavy fins, moves with an insinuating glide. The stout muscles under the rough skin give uncanny force to his onward drive. The leering, chinless face, the evil, murderous, underhanging jaw, the cold staring eyes, the gleaming, jagged teeth impart to the man eating shark that diabolical aspect which has made him the terror of human mind on all the seas. Like most of the tribe, he turns himself on his back to bite, for as his underjaw is so much shorter than the upper he must do this to keep the snout from interference. When he sees a small object on the surface he will aim for it in the natural position and lift his head out of the water to snap for it.

The structure of the white shark has been studied from a few specimens which have been obtained in the southern waters of the United States. A very good specimen was taken in December, 1913, by Sidney M. Colgate of this city and his party, who went out on a cruise from Palm Beach with Bert Hisock, an expert fisherman, as their guide. It is common to go shark fishing down in that region, where there are many of the so-called ground variety which attain great size.

The party harpooned a shark that at once put up a strong fight. First he bit the wooden shank of the harpoon in two and then started for the boat. He had already ripped off one of the upper planks with his sharp and powerful teeth when several revolver shots taking effect near his head put him out of commission for good.

This shark, which had all the characteristics of the man eater, was hauled up on the beach and a series of photographs made, which have been of value to science. The serrated teeth are shown by these studies to have that peculiar sharpness possessed by those which killed young Bruder at Spring Lake.

The jaw of the shark, even of the large man-eater, is not as strong as it is commonly supposed to be, and few except the larger ones can bite through the bone. Where the legs of human beings are taken off the shark fastens his teeth into the limb directly over a joint, so that the amputation is made through the tendons. As the great fish can drag his victims under water, and so drown them, he can dispose of them at his leisure.

Dr. Charles H. Townsend, the director of the New York Aquarium, who has observed the habits of fish in all parts of the world, tells of the greedy habits of the shark as he studied them which he was making the fisheries survey on the *Albatross*. The sailors, partly for amusement and partly because these sons of the seas have an inborn desire to avenge themselves on sharks whenever they can, used to angle for them from alongside, haul them up on deck with tackle and kill them. The contents of the stomachs of these ocean scavengers represented everything which could possibly be thrown over from a ship. The great white shark is the prize glutton of the deep. In one specimen were found the hindquarters of a pig, the forelegs and head of a bulldog, some horseflesh, half bushel of mutton bones and several tin pails.

Even in the tropical zones the causalities due to sharks are comparatively few, considering the great risk when the natives will frequently take in the water. Dr. Townsend spoke the other day of one or two cases in the Society Islands, in which natives who were riding on the ponderous surf planks, the forernners of our aquaplanes, had fallen off and were dragged under by sharks. In the Far East where thousands of men make a living by diving for the pearl oyster, sharks are seen in large numbers. Many of the creatures are killed by deft knife thrusts of the divers.

The people of the West Indies are at many of the ports almost amphibious. They have the dreaded man eater, without sustaining injuries themselves. Several years ago by means of an ingeniously constructed diving bell apparatus reached from a collapsible tube, pictures were taken at Nassau of the life under the sea. There the moving picture operator who was working for the Universal Film Company took the so-called Williamson reels, which showed the dark skinned divers actually killing monster sharks. As the shark must turn on his back to make his bite effective, the diver could swim about him, grab a fin and send a fatal stroke home just at a vital spot.

Although the natives have an almost superstitious fear of the tiger shark, so-called from his stripes, the cases of deaths from sharks of all types are almost as rare in the Antilles as they are on the seaboard of the United States.

The fight between man and shark is a hard one for the big toothed fish has wonderful vitality and insensibility to ordinary pain. After a whale has been harpooned and taken alongside sharks frequently gather in schools waiting for the cutting to begin. If there is delay they will bite into the whale on their own account.

The sailors who are taking off the sheets of fat will try to drive the sharks away by striking them with sharp blubber spades. So callous to suffering and so greedy are the sharks, however, that even after they have been mutilated they will return again and again to attack the carcass.

So much for the fierce and predatory sharks, but after all even these tough skinned vagabonds of the deep have a tender side. They often travel with several varieties of fish who are their devoted friends. There is the remora that clings to Sir Shark by a sucking disk, and where he goes Little Remora goes. Indeed, he sticks to him closer than a brother.

Then there is the little striped pilot fish, which swims along with the shark or in front of him, as though to show the way and give him timely warning. It is recorded that once when a shark was hauled on board a vessel hooks were cast over for the seven pilot fish in his suite, and such was their devotion that one by one they permitted themselves to be caught and hoisted aboard to join him in his funeral.

Sharks are hated so much that few there are who appreciate any of their good qualities. Some of the smaller kinds, with their firm white flesh, are good to eat. The Chinese esteem the shark fins as a great delicacy. Large quantities of the fins are shipped to the Celestian republic from our Western coasts.

Shark oil was in high request as a lubricant for watches. The skin of the shark and of fish of allied species is used in the making of shagreen, a leather which, especially in the hands of the Oriental artificers, can be made into decorative leathern cushions and book and manuscript covers of rare beauty.

And wild is the nature of sharks, think not of them without their family ties. In the summer, when the Gulf Stream glows, they drift into the bays of the New York and New Jersey coast. There the young are brought forth alive, and are often caught by fishing parties. At such places as Long Beach are found the dogfish, smooth skinned and agile, and also the young hammerhead, a type between a T square and a semaphore, with his eyes perched out at the end of bony crosspieces. These babies often grow to enormous size, and even the hammerhead has been known to reach a length of thirty feet.

You who have visited Beach Haven and Egg Harbor and such resorts will probably remember the venerable fisherfolk who took you out in catboats for the express purpose of killing "sheards," as they call them, and many a young might-have-been monster has been dragged wriggling from the blue.

Let's not think that the shark is an exotic, as far as we are concerned, along these coasts. Often their shining backs may be seen off Robin's Reef and the glint of their fins is spied off Bayonne despite the sludge which is so hard on placing wanderers.

Sharks have been caught for many years off the Jersey shore. The grizzled fishermen find many a fine specimen in the big pound nets near Sandy Hook. In fact, most of the injuries which men have received from the teeth of the tails of sharks in this vicinity have been due to their efforts to kill the creatures with clubs or axes after they have been caught.

As a general thing the shark is cowardly and is not looking for a fight. Only the great white one has the name of spoiling for trouble. Many a sailor no doubt has been devoured by sharks in our latitudes, but only after life had long been extinct. The favorite food of sharks in these waters is fish, especially flounders or plaice, creatures which lie close to the bottom. Fire Island inlet is a favorite hunting ground for the tribe.

The common type of shark found about here is the ground shark, which attains a length of from six to ten feet. It can be found in the southern waters and to many points along the Atlantic as far north as Cape Cod. The small toothed nurse shark is frequently seen by the fisherman, and there is a record of one being brought ashore in 1885 by the crew of the Amagansett (L. I.) Life Saving Station.

The tiger shark, like the man-eater, is a straggler, and has been seen along the Massachusetts coast. One was killed off Long Island in 1911 by Captain John Doxsee. Two small man eaters were taken at Wood's Hole in 1903.

The great blue shark, which is one of the hardiest specimens and a special friend of little Brother Remora, is one of the finest looking of the genus. The dusky shark is a peculiar creature which the New Jersey fishermen call the "Santiago" because there is a tradition that his shocked family left the coast of Cuba on account of cannonading which ensued when the fleets of Cervera and Sampson met. He is also called the Spanish shark for they of the smacks have a way of naming every fish Spanish whose looks and habits are mysterious to them. The shark is thoroughly at home in the Great South Bay, and inhabitants of Babylon often go in quest of them with big hooks and salt pork.

There are many other varieties which have been described by the local naturalists, such as the spotted fin, the round nosed, the sharp nosed, the shovel head, the thresher, the sand shark, the porbeagle and the basking, while the dreaded man eater or great white shark is related to the mackerel sharks, so called on account of their ship-shape lines, which are much like those of the mackerel.

The shark being no stranger to us, and not nearly as dangerous as his reputation indicates, should none the less not be considered as anything like an old friend.

Dr. Frederic A. Lucas, director of the American Museum of National History, says that an attack by a man eater shark in this part of the globe, which is apparently well authenticated, is rare, indeed without precedent. He compares it to the sudden bolt from the blue. He long worked along the lines suggested by the reward offered by Mr. Oelrichs, and had been unable to substantiate the stories of loss of life.

John Treadwell Nichols of the department of ichthyology of the museum said yesterday that the accidents off the

Jersey shore were so unusual that it might be possible that both victims were attacked by one shark. He declared that the man eater was much an unaccustomed visitor that the likelihood of a bather's being injured was about the same as he might someday be struck by lightning.

The scientists believe that taking everything into consideration the chances of persons being injured by sharks are slight if ordinary precautions are observed.

From the *BROOKLYN DAILY EAGLE*,
Friday, July 14, 1916

*Long before every newspaper story had to have a graphic accompaniment called a sidebar or Infographic pictures,* the Brooklyn Daily Eagle *published these facts of the times to accompany their shark coverage:*

## FACTS ABOUT SHARKS

- There are about 150 different species.
- The shark's scientific name is Plagiospond, sub-order Squali.
- Sharks are most numerous in the tropics.
- Some live at a depth of 1,000 fathoms.
- The sharks common on the coast of England have teeth, and the female brings forth some thirty living young at one birth.
- The tiger shark is common to the Indian Ocean.
- It grows from 10 to 15 feet long.

- Nineteen species of sharks have been found in the water of Long Island.
- Of these, only the man-eater and tiger shark could have killed men.
- Extinct species attain a size of 90 feet.
- The basking shark is the biggest of the present species, and reaches 20 feet in length.
- The species never attack human beings.

# HMS VALERIAN

*L*eaving *Nassau in the Bahamas headed for Bermuda,
HMS* Valerian *ran straight through the eye of a hurricane and plunged into the raging storm on Saturday,
October 22, 1926. Ironically, the ship had inspected
hurricane damage with the governor of the Bahamas
earlier, touring outlying islands to survey the damage.
Now she was in the thick of the storm, with sustained
winds of ninety-five miles per hour and gusts reaching 136 miles per hour. A 1,250-ton sloop, the ship was
initially used as a mine sweeper and had triple hulls
at the bow to protect against potential mine discoveries of the worst kind. But it was not built, nor had the
size, to withstand hurricane-force winds. The* Valerian
*foundered and around 1 p.m. on October 22 it eventually went over in heavy seas. At that point, eighty survivors made it into rafts. According to the newspaper
report below, two people were snatched off the raft by
sharks. Only nineteen survivors were ultimately rescued
on October 23 by the cruiser* Capetown, *including the*

*commanding officer and navigating officer. There are no reports of the death toll from sharks, but surely more than two passengers were killed. Drowning victims were food for the sharks, too, of course.*

*The newspaper report below from 1926 also gives some information on the sinking of the* Eastway *in the hurricane.*

From the *Evening Journal* (Wilmington, Delaware),
Friday, October 29, 1926

## SHARKS PULLED MEN OFF RAFTS

Surrounded Members of Crew After the *Valerian* Had Foundered.

## TWO SAILORS WERE DRAGGED INTO SEA

*NEW YORK,* Oct. 29—Sharks surrounded the rafts from which the nineteen survivors of the British sloop of war *Valerian* were picked up last Saturday, according to passengers arriving from Bermuda—yesterday aboard the *Fort Victoria* of the Furness Bermuda Line. They brought fragmentary accounts of the foundering of the British patrol vessel with its loss of eighty-four lives, and also the story told by the twelve survivors of the British freighter *Eastway*, which sank in the hurricane last Friday night with the loss of the captain, two officers and twenty seamen.

Eighty of the crew of the *Valerian* were aboard rafts after the vessel foundered, survivors reported. Two were seen pulled off by sharks and others dropped off in the high seas until only nineteen were found twenty-four hours later when the *Capetown* reached the scene, about eighteen miles southwest of Bermuda. The *Capetown*'s officers reported they never had seen the water so full of sharks.

Commander Usher, Flag Lieutenant Hughes and the seventeen other survivors are under military arrest, charged with losing their ship, and will be so held until the official hearing. They rode out the first blow within sight of land when the wind reached a velocity of ninety-five miles an hour. A dead calm followed and then the second curve of the hurricane struck the sloop, a 1,250-ton vessel, amidships and she went over.

Survivors of the *Eastway* said they were 150 miles northeast of Bermuda when the gale struck. Their main hatch was stove in and all but one of the lifeboats smashed.

Captain Van Stone left the bridge about 4 o'clock and went on deck to try to cover the hatch through which the sea was pouring. A wave caught him just as he reached the deck and washed him over.

By 6 o'clock that night the coal in the bunkers had shifted and the first S. O. S. was sent out. It was answered both by the *Fort St. George* and the *Luciline*, a British tanker, bound for New Orleans. As the tanker was about thirty miles away, the *Fort St. George* let her go to the rescue.

The *Luciline* reached the positon given and cruised about all night and until 10 o'clock Saturday morning, when she saw the sail in the lifeboat of the survivors just as she was staring away. She turned back and an hour and a half later had the men aboard.

They said the first and second officers went down on the bridge and R. James, the radio operator, in his cabin, where they heard him calling for help within five minutes of the time the *Eastway* turned over and sank. Just as the steamer went down, two apprentice boys jumped overboard and were not seen again. One of the Arab firemen cut the last lifeboat loose just as the steamer turned and into this the survivors climbed.

They reported that half an hour later a large steamer passed them so close they could see her deck lights. They sent up seven flares but the steamer passed on. Captain Thompson of the *Luciline* told them he saw her later, but she refused to answer the *Luciline*'s calls.

The twelve survivors of the *Eastway* will be brought to New York Monday on the *Fort St. George*. Four are English seamen and the other Arab firemen.

During the hurricane at Bermuda, the *Calcutta*, the flagship of the British West Indies squadron, had a close call from pounding to pieces against the dockyard jetty. She was moored with twenty-eight hawsers, had one anchor down and steam up. When the gale hit, all but one of the hawsers snapped and the ship was flung against a pierhead. Two officers leaped off the jetty and swam down the lines which were hauled aboard. After a nine-hour fight with wind and sea the vessel was saved.

Damage at Bermuda was estimated at $500,000.

# PRINCIPESSA MAFALDA

On October 25, 1927, the Italian transatlantic ocean-liner SS *Principessa Mafalda* was bound for Buenos Aires when she went down with 1,252 passengers and crew aboard. The disaster was the greatest tragedy in the history of Italian shipping, as the ship was operated by Navigazione Generale Italiana. With a final death toll of 314, this was also the worst peacetime loss of life in the Southern Hemisphere. The sinking of the *Titanic* in 1912 happened in the Northern Hemisphere, with more than 1,500 dead, and the fate of the *Mafalda* drew comparisons. Perhaps a more accurate, and understandable, comparison is the more recent sinking of the *Costa Concordia* off Italy's Isola del Giglio in 2012. Both ships were luxury cruisers of their time on pleasure voyages, though the *Mafalda* was also transporting overseas passengers and therefore had different passenger classes booked aboard—first class to third class or steerage. The reported incompetence of the crews on both ships that went down, and the tumult that took place during and after the wrecks, also invited comparisons. The major differences in the disasters were the overall loss of life—more than 300 on the *Mafalda* versus thirty-three on the *Costa Condordia*—and, more horrific, the presence of sharks at the 1927 wreck.

Originally, newspapers reported that 470 died out of the 968 passengers and 230 crew (an estimate of 1,198 total) on the *Mafalda*, after an initial overreporting of 1,600 aboard the vessel. (Extending the comparison mentioned above, the *Costa Concordia* had more than 4,000 people aboard in 2012, so thirty-three lost seems a surprisingly low number. Modern safeguards and rescue techniques no doubt minimized the severity of the tragedy.) A United Press report on October 26, 1927, said that 720 people were saved and brought to Rio de Janeiro. (Variance in total numbers of passengers and the resulting dead and survivors was not unusual back then, as newspapers were the main source of information and the spread of news and updates were slow and fact-checking was not always possible.) A 1927 report from United Press read: "The fact that when she sank she had reach only the vicinity of Bahia—a day's steaming from Rio—yesterday evening was taken as certain evidence by officials that something had happened to her far at sea to delay her progress." On October 23, survivors reported that the ship had to stop repeatedly due to mechanical problems; the ship also began to list, survivors later reported. On October 25, several strong shudders were felt passing through the ship, possibly from a propeller coming loose and striking the hull.

History shows that the *Mafalda* suffered a broken prop shaft, which allowed water to rush into the engine room. The water eventually flooded the boilers, and the foundering ship—the crew issued an SOS distress call at 7 p.m.—sank just after 9 p.m. on October 25. This occurred in a well-traveled shipping lane, and several vessels responded to the SOS call. Records show that a multinational brigade picked up survivors: the Dutch steamer *Alhena* rescued 450; the British steamer *Empire Star* picked up 299; the German steamer *Baden* brought aboard seventeen; and the Italian steamer *Rosetti II* rescued eleven. The other ships mentioned in the rescue were the *Formosa* and *Mosella*. All survivors were brought to Brazilian ports. Other vessels may have been diverted to the scene by "forced draft" but did not engage in rescues or did not report appreciable numbers of survivors picked up.

Survivors said "huge, fierce sharks" were attacking people in the water. A survivor picked up by the ship *Mosella*, which brought about fifty survivors to Bahia, Brazil, said he saw the terrifying image of sharks darting in and out from the darkness into a "circle of light" where spotlights were aimed to locate survivors. He said one steward of the *Mosella* saved a man by shining the beam of a flashlight into the eyes of a shark about to attack the man. An Associated Press report read, "Some of [the survivors] insisted that a number of victims were devoured by sharks as they struggled in the water."

There was no way for officials to know how many in the water succumbed to the sharks, versus drowning or dying of exposure or exhaustion; several reports say passengers died from exposure, despite the Southern equatorial location. But eyewitness testimony included descriptions of terrifying and fatal shark attacks and "a horror-filled night" after the steamer went down. Another report after the incident read, "The rescued men appeared much shaken by the sinking of their vessel, some of them even declared that sharks devoured victims of the disaster."

The ship was an accident waiting to happen, it would seem. It was running off schedule throughout its voyage from Genoa, and after leaving Barcelona, Spain, the ship was clearly laboring due to mechanical problems. A report in the *El Paso Herald* of El Paso, Texas, quoted a first-class passenger named George Grenade who said he sent a letter to the Italian royal maritime commissioner with his testimony that, "The commander and first engineer confirmed that this was to be the *Mafalda*'s last South American trip. She was to be sold for Mediterranean cruises for her condition did not [*sic*] longer permit long voyages." In fact, it was reported that when the ship reached Cape Verde, the captain requested by telegraph that another ship be sent to replace the *Mafalda*, but his request was denied or ignored.

The *Mafalda* apparently did not have lifeboats ready to launch or they were improperly placed, and perhaps not enough lifeboats were available. The hour of night, and darkness, did not help matters for those spilled into the water. It was described as chaotic—understandably so—though passengers were said to be panicking and looting valuables. The ugly side of humanity flared up as tragedy became reality, and the sharks equalized the situation by eating people regardless of their comportment or violent or unruly behavior.

News reports at the time said Captain Simon Gulî drowned and the chief engineer took his own life. Now, if you are superstitious, the name of the ship might be a harbinger that disaster was inevitable: Princess Mafalda of

Savoy, the daughter of King Victor Emmanuel II, died at Buchenwald in 1944 while imprisoned there by German Nazis in World War II. After the *Mafalda* sunk, it was often described as "ill-fated" in news reports. It had been one of the prettiest and fastest ships in the Navigazione Generale Italiana fleet for years, and with two-story ballrooms and rooms designed in the Louis XVI style, it was a luxurious if not opulent vessel. However, by 1926, it seems her beauty had faded and her body had weakened and began failing, so much so that she was likely done in by her own propeller and faulty propeller shaft. And the sharks were waiting.

*Imagine: You are running along the ship's upper deck, desperately searching for open space on a life boat, your mind is raging with internal argument, at the same time pleading to stay calm and imploring with your body not to panic, when a surge of malevolent, unthinking, enraged, and terrified passengers mobs you and one burly man pushes you over the railing of the listing ship. You flip in the air and your back slams onto the water, which strikes you as surprisingly warm and bath-like, and you come to the surface in a scrum of bodies, some thrashing and others hulks of wet clothes, lifeless. "Thank God I didn't land on anyone," you think as your head clears. A beam of light sweeps over you and you see a torso bobbing on the surface chop; the night is calm, the sky is clear, you hear screaming and splashing, and distant screams from above, people still on the ship. Something bumps your legs, and you swirl and kick, reflexively. The person in front of you suddenly surges up in the water and shrieks and screams, beating the water with his fists, and is pulled under, his head snapping back as his body jerks under the surface and is gone. An hour ago you were sipping a highball in the ballroom, watching your sister as she danced with a man she met earlier in the day on the upper deck—he was in imports, didn't he say? Something about coffee or was it sugar cane? Or bananas? A body pops up on the surface, the same jacket as the man who just disappeared. His legs are gone, it's only a torso bobbing in the waves. "Hey!" you yell . . . and then you scream and as a hot heat seizes your leg and tugs you once, then again, and you're pulled under through the swirling bubbles and you see a flash of the sweeping beam of light, and a final thought dawns on you—"It's the spotlight and they're searching for survivors . . ."*

# RMS *NOVA SCOTIA*

In 1942, as World War II warfare raged in and around Stalingrad in Russia and attacking German soldiers were dying at a rate of 10,000 per day, the Russians continued their homeland defense through the cold and snow. It was a battle of sharks versus sharks.

"While clouds vanished, cold prevailed in the Stalingrad area. Wind-driven snow cut down visibility on the central front. Moscow observers said the joint operations were the greatest undertaken by the Russians in their 17 months of war," read a report by the Associated Press wire service. Stalingrad became known as the city that defeated the Third Reich, as the crushing defeat of the German forces by the Red Army in urban warfare was a pivotal loss for Germany. After six months of horrific battle and a death toll of perhaps more than a million, the city became known as "Red Verdun," in reference to the horrible battle of World War I on the Western Front between Germany and France, which resulted in almost one million dead during the battle, and 1.2 million dead at Verdun throughout the war. The Battle of Verdun lasted for 303 days and was the longest and one of the most brutal battles in the history of the world. A Dresden newspaper called Stalingrad "the most fateful battle" of World War II. Winston Churchill wrote of Stalingrad in his history of World War II titled *The Hinge of Fate*: "The cold was intense: food and ammunition were scarce and an outbreak of typhus added to the miseries of the men." The Germans and Field-Marshal Paulus and the fierce German Sixth Army surrendered the city in February 1943. Russian Premier Joseph Stalin called it "the annihilation of the encircled enemy troops near Stalingrad."

While that battle was raging in Europe in autumn 1942, the German navy had a well-documented conquest, off the coast of Africa. The 6,796-ton RMS *Nova Scotia* had been requisitioned by Britain as a troopship and in 1941 joined a convoy sailing for Sierra Leone. Now near the African continent and recently departed from South Africa, the *Nova Scotia* in autumn 1942 was deployed to the British Colony of Aden with 750 Italian prisoners of war aboard, bound for Durban. However, off the coast of Natal Province, South Africa, a German submarine intercepted the ship with a volley of three torpedoes.

The *Nova Scotia* went down quickly, one report said within ten minutes. Not much is known about the aftermath, except mass death occurred. Of the 750 Italian prisoners aboard, 650 were reported lost. The total death count was 858. Many drowned, relatively few were rescued—fewer than 200—and corpses were reported floating in the sea and washing ashore.

In its time of service, the German submarine *U-177* sank fourteen ships and was later sunk by US depth charges, with the loss of fifty crewmembers and fifteen survivors from that incident. A peculiar fact of the sink-ing of the *Nova Scotia* was that *U-177* picked up two sur-vivors to find out what ship they had hit. They were Italian detainees who told the Germans that most of those aboard were also internees. The Germans were under orders *not* to pick up enemy survivors, as that had occurred follow-ing the sinking of the RMS *Laconia* and US warplanes attacked the German U-boat, forcing it to crash dive, kill-ing survivors on the submarine's deck who weren't strafed by the US gunfire. This brought unrestricted submarine warfare to World War II, since humanitarian acts were now against orders. The German *U-177* commander Robert Gysae reportedly left the two survivors from the *Nova Scotia* in the water and continued on. I found no reports on their eventual welfare, but if oceanic whitetips were in the area we can surmise the outcome.

Those in the water after the *Nova Scotia* went down had little chance of survival. It wasn't long before oce-anic whitetip sharks showed up on the scene, as they did at so many wartime tragedies. These were, indeed, shark-infested waters, and oceanic whitetips—which the undersea explorer Jacques Cousteau called "lord of the longhands" because of their pronounced and white-tipped and wing-like pectoral fins—came calling. Once again, as was so common on the seas in World War II, it was feeding time for the sharks.

# THE PIG BASKET ATROCITY

By Stephen H. Foreman

*The brutality of war is not driven by human instinct, it is conjured by the ugly will of man to send terrifying messages and examples of horror that will force the enemy* du jour *to submit. In a way, the intent is altruistic: Making the enemy quit will curtail the overall deaths; the sooner the battles come to an end, the sooner human existence can continue—with the victors now free to decide what that existence will be. War engenders unspeakable horror, a fall from grace so irretrievable as to make God and His angels hide their faces in despair and mourn their creation. In World War II, at times, sharks became the method by which the horrific was so heartlessly achieved. History has written these events in blood. For writers, mere mortals, words are insufficient, although the need is to try with what few means of expression we have. The Japanese conjured such depraved methods—including the unholy incident of the* Pig Basket Atrocity.

To kill in war means engaging the underbelly of the most feral aspect of the self, to slaughter without feeling, to rip apart the enemy with no mercy, to become inhuman—to attack, in other words, like a shark.

The terror wreaked by Japanese soldiers on both military and Chinese civilians when they invaded Nanjing (also called Nanking) in the Republic of China was so horrifying as to be nearly unfathomable. History knows it as "The Rape of Nanjing." From December, 1937, to January, 1938, this one city in China had a death toll estimated to be more than the dead of the two atom bombs dropped on Hiroshima and Nagasaki—men, women, children, the elderly; a reliable estimate totals somewhere in the range of between 200,000 to 300,000 murdered. The operative strike was in the form of instant, unprovoked death, multiple rapes perpetrated on women of every age (as young as twelve, as old as seventy), and ghastly mutilation. People were herded into houses that were then set on fire. If they tried to escape the flames, they were gunned down as they ran out the door. Civilians were bayoneted, and the dead were used for bayonet practice. Babies were ripped from the bellies of pregnant women. Officers tested the sharpness of their swords by random decapitation. There was no thought given to this. The victims had only to be standing nearby when an officer decided to behead them.

There was little difference between shark attacks and the bloodbaths executed by the Japanese soldiers, the difference being that the Japanese soldiers were more imaginative. They even went so far as to make sharks their instruments of death. Of course, sharks are not imaginative or, in the human definition of the word, cruel. They don't know or care if you are Jewish, Black, Chinese, Caucasian, or a pig that had the misfortune of falling into the water. All these creatures know is that food is up there on the water's surface. They have no thought process, *no* concept of intention. Sharks simply act in the manner in which nature has programmed them. Human beings do have a thought process, intention, yet the terror inflicted by one man on another, the mortal dread, the blood-cold panic, the total absence of mercy—these were the common denominators. And so we have the Pig Basket Atrocities. However, even before *that* we have the German death camps and the Japanese Unit 731.

History attests to the fact that during World War II, Josef Mengele's "medical" experiments at Auschwitz were notorious for the suffering coldly inflicted on people the Nazis deemed barely human. Mengele was known as "The Angel of Death." One of his survivors said, "If there is a devil, it lives inside Josef Mengele." Another said, "I was only alive as Josef Mengele wanted me alive."

Imagine lying alone on a cold steel lab table watching as a man, calm and collected, nattily dressed, the object of deference from everyone around him, approached you with a rusty scalpel in his hand, knowing not what he was about to do but surely that you were seconds away from unimaginable pain. When asked why she didn't try to escape, the survivor replied, "There was no place to go. They could do whatever they wanted." And again, "I had difficulty coping with the fact that I was a nobody and a nothing—just a mass of cells to be studied." Say it another way: "Just a hunk of food to be consumed."

Understand, this is not an attempt to demonize sharks for doing what they have naturally evolved to do. Since the debut of the movie *Jaws*, such a cursed public image has infected the entire world with an unwarranted yet visceral sense of dread, resulting in the needless slaughter of literally millions of sharks, though many have also turned to a conservation-minded path, including the author of the novel, Peter Benchley. I would be remiss if I didn't point out that in our own enlightened country, the United States of America, there were also people once deemed barely human. They were kidnapped from Africa and called slaves.

And what could it have felt like for our sailors, soldiers, and airmen shot down or torpedoed during World War II, treading water in the ocean, knowing that under the fins circling around them were massive, murderous engines in search of food? Death machines that didn't care or even comprehend what their victims were, even what victims *were*, guided by sheer instinct, knowing nothing, not even that they didn't care, not even that they had no concept of care at all.

Please allow me to digress a bit into personal history. When I first set foot in Hollywood as a screenwriter decades ago, my mentor, a Major Shark, explained it this way: Hollywood is a giant shark pit. On the bottom level there are a bunch of little sharks who eat each other. On the next level exist the midsized sharks that eat one another as well as all the lesser ones. Then there is the top level—the great whites. They eat everybody. A conscience? What's that?

I happened to become friendly with another great white, a studio chief. I don't know why he liked me, but he did. In turn, he fascinated me. One day I was in his office when the door banged open and a furious producer rushed in. "Goddamn it," he yelled at the studio chief, "Yesterday you told me such and such, and this morning I find out it wasn't true!" The guy was apoplectic, definitely on target for a stroke. The studio chief remained as tranquil as Buddha under the bodhi tree, stared the poor fellow right in the eyes, and said, "I lied. So what? What're you gonna do about it?" Dante put it best when writing about Hell, "Abandon all hope ye who enter here."

The Japanese version of Hell was Unit 731, truly another pit of horror, torturers and victims different only in skin color from their counterparts at Auschwitz. Indonesian and Filipino civilian populations were dissected alive without anesthetic because doctors wanted to see the effects of introduced pathogens on the organs without them becoming tainted by killing off the pain. Limbs were amputated and grafted to opposite sides of the body to see the effect of gangrene. Young women, aged eighteen and nineteen, had their wombs sliced open with scalpels as a lesson in sex education designed to instruct the younger soldiers who knew "very little about women." When asked why, the reply was, "It was the Emperor's orders, and the Emperor was a god." (A crystal-clear example of why our Founding Fathers, in their wisdom, enshrined the need for separation of church and state in the new government of the United State of America.)

At Unit 731 prisoners were staked outside in bitter cold with ice water continually poured over an exposed arm to ascertain the length of time it took for frostbite to set in. To test whether or not it had, the arm was hit with a stick, and, if it made a hard, hollow ring, the limb was considered frostbitten. Others were put into decompression chambers to see how long it would take for their eyeballs to explode. Russian, Filipino, and allies were infected with anthrax, typhoid, dysentery, and cholera, and then pickled in formaldehyde. Liberating troops discovered a six-foot-tall glass jar holding the remains of a grown man. Yet, all these crimes went unpunished because our government provided immunity to the "researchers" in exchange for the results of their research. Could anything have been more horrid? Did knowledge obtained so unscrupulously trump the human heart?

Now, let's talk about the Pig Basket Atrocities.

A pig basket is an oblong affair, three-feet long, made of bamboo, designed to hold pigs for transport. Ironically, it has the shape of a cornucopia, a horn of plenty—wide at one end, narrow at the other. In ancient times, in a practice called *zhu long*, couples convicted of adultery were put into pig baskets—one for the male, one for the female— and then dropped into deep water and drowned.

During the Japanese invasion of Nanjing, a troop transport was torpedoed and sunk. Friendly fire was suspected. The general, Hitoshi Imamura, was forced to swim ashore, a singularly demoralizing experience guaranteed to leave an officer of his rank and status in a foul mood. The irony of this man, however, is that he was unusually lenient toward the local residents of the Dutch East Indies, an attitude that led him into conflict with his military senior staff. Nonetheless, he persisted, and his insistence on continuing with this policy minimized greatly problems with the occupation, and yet, only two years later, General Imamura was responsible for the Pig Basket Atrocity.

When the Japanese conquered East Java, two hundred Australian and British soldiers took to the hills and regrouped. From there they waged guerrilla warfare against the occupying forces. Although vastly outnumbered, they fought valiantly and held out far longer than seemed possible. Eventually, however, the Japanese hunted them down and captured them alive. Ten were immediately bayoneted to death. The rest were denied food and water, forced to stand in the hot, tropical sun for days. When their captors deemed these men weak enough, the prisoners were trussed hand and foot and forced into pig baskets, for the most part two per basket, one top, one bottom, face to face, under 100 degrees of broiling sun.

Civilian residents who survived as eye witnesses remembered the horror of these men screaming from cramps, crying, begging for water. They remember attempting to bring water to the prisoners only to have soldiers with bayonets prod them to keep them from getting close. Appalled neighbors tell of an elderly woman who managed to evade the guards with water for a dying prisoner. A quick flash of a sharp sword, and her severed head was left to rot in the sun.

Many years later, an elderly woman who observed this event as a child continued to hear the howls of pain and madness, the heart-wrenching cries for mercy until her death many years after the war. She saw a guard unzip his pants and urinate on the prisoners, and she watched as the pig baskets were loaded onto open trucks and transported to a railroad siding. There the pig baskets with their human cargo, nearly 200 men, were stacked onto flat, open cars and taken to a ship anchored off the coast of Suribaya, where they were transferred to its deck. The ship then put out to sea, and, when it reached shark-infested waters, the prisoners in their bamboo baskets were pitched overboard to die—to drown or be eaten—if they were not already dead.

No doubt these were the whitetip sharks so prevalent in deep-ocean and temperate water, the very species that likely accounts for more human kills than any other species in history, many more than the dreaded bull shark and certainly the great white. These are pelagic creatures, which means they only live in deep oceans, 150 to 450 feet below the surface. They rarely come near shore, which is why they are not so well known. Their attacks take place in the open ocean on the "survivors" of ships that have sunk and planes that have been shot down. At first there is only one shark, and then there are many, snapping, biting, attacking, and feeding in a frenzy. Throughout World War II, Japanese submarines and kamikazes made certain the whitetip sharks did not go hungry.

Imagine being bound and crammed into a bamboo basket, trussed like a pork loin straight from the butcher shop, and, if you were still alive, staring as the gaping jaws and knife-like teeth of a ten-foot-long predator weighing hundreds of pounds zeroes in on you. A man's body weight would eventually sink the basket, and suspended in the surface zone the basket would be a tempting target for the host of aroused and fevered sharks. Maybe the weakened soldier would drown before the shark smashed the bamboo like so many toothpicks and the cloud of human blood spurred it and others on to feast on flesh.

Again, this was not a spur-of-the-moment punishment. Someone had to sit at his desk and think this one up. That someone was Lieutenant General Hitoshi Imamura.

Imamura's was the benign face of a favorite uncle or a beloved grandfather, shaped like a melon or a ripe spaghetti squash, yet the man's behavior was that of a sadistic killer, uncaring, with no conscience, ruthless, hungry to conquer, relentless in his desire to destroy in the cruelest way possible, without a shred of empathy. Some might say this was shark-like behavior. Negative. A shark may be hungry, but its sole purpose is to eat, not to conquer, never to intentionally inflict pain.

After the war, Imamura was convicted and served a ten-year prison sentence in Sugamo, Japan. His victims, were they alive, probably would have preferred he be crammed into his own custom-made pig basket and, very much alert, dumped overboard into very deep, very temperate, very whitetip shark–infested waters.

Now, visualize the following: from contemporary, state-of-the-art, underwater video of uncanny clarity, we've seen for ourselves how great white sharks reach a frenzy while biting their way through metal shark cages containing cowering tourists who paid to see sharks in their element. These primordial creatures can actually destroy a metal cage specifically designed and constructed to keep them out! They open their mouths, brandish those teeth, and charge. You see the caged humans instinctively leap backwards out of righteous terror. And again the sharks charge and batter the cage, this time breaching further, their heads penetrating within scant inches of a man's face, near enough for the man to touch. There can be no doubt in the caged human's mind that he is being attacked by a creature determined to tear him to shreds and devour him. One man had to throw himself on his back to escape a shark that had penetrated so deeply into the cage it nearly got him. Again, the shark charges—it is relentless in its attack—and eventually gets through the bars. The man I observed escaped by wriggling through the exact opening made by his attacker. He and his predator were actually side by side, only the shark was too intent on attacking the cage that it, for whatever reason, did not turn on the man. Imagine what he must have felt being in such dangerous water, his attacker's natural habitat, trying and hoping to swim to safety?

There is aerial video footage taken from a helicopter of a summer beach day with hundreds of folks—families, the elderly—frolicking in the surf along a well-traveled vacation beach. Frolicking with them, swimming all around them, sharing the water, unseen and unknown, are sharks, many of them bull sharks, the species scientifically known for attacking human beings more than any other. There are also a number of underwater videos showing tourists romping with sharks, feeding them, smiling at the camera. Has anyone ever seen a shark smiling back? After scanning dozens of shark videos, I sure never have, and I'm sure as hell not going to get in the water with them, either (I'm even wary of swimming pools these days). My feeling is that this is lunacy. Maybe with dogs, maybe with horses, we share a connection—however, with most wild animals, no matter what a person fantasizes or believes, we are not kindred spirits. One can observe and know animals as only an observant human being can. Good deer hunters can strive to "become" their prey as they look for it. Their senses are attuned with the wind, with every step taken, but there are no ties of blood. They are not kin, certainly even less so with a fish than with a mammal, but even then we are not consanguineous or, to put it another way, collateral relatives.

Take Timothy Treadwell, for example. Treadwell spent years camping in grizzly territory in Alaska. In all that time, the man had no problem with the grizzlies—a number of which were always around. He became familiar with the bears, and they became familiar with him. He watched them from within their critical distance for a multitude of hours, especially one in particular, an immense male, 1,000 pounds, nearly ten feet tall when standing on his hind legs, a titan nearly twice as tall as the man, towering over him. When the bear was on all fours, he stared Treadway right smack in the man's face. Any bear is a force, but one like this is a howling hurricane force, and, no matter what Treadway thought, like a hurricane, not to be trusted. Treadway, of course, thought differently. He even brought his girlfriend out there with him. She was a photographer and took pictures of Timothy and his ursine *friends*.

The last anyone ever saw of Timothy Treadwell and his girlfriend were their remains in the giant bear's stomach. A sound recording exists of the surprise attack with Treadwell screaming that the bear was killing him. A recording device happened to be on. His girlfriend watched the assault until Treadwell was dead. Then the grizzly turned and killed her, too.

No, dangerous animals are not propelled by vengeance or diabolical motives. They kill because they have to eat. Treadwell suffered the inevitable end that his constant proximity to wild grizzly bears ensured. Too often, so-called fearless adventurers take calculated risks that end with the animal doing what it instinctually does—kill what it then eats.

Shark attacks have increased in recent years. Why? Because we humans have breached their territory, entered into and modified their domain: by over-fishing their food supply, by hunting them for dorsal fins marketed as an ingredient for soup, by a tourist industry that beguiles its customers with unpredictable underwater encounters (such as the shark-cage episode described earlier), by a tourist industry that lures innocent yet ignorant customers to surf and swim in shark-infested waters.

The incontestable evidence is that animals in the wild are afraid of human beings. We stink of death. We carry it with us. That's why they flee at the first sound, scent, or sight of us rather than seek us out—except the shark, a predatory creature that instills in us the same terror experienced by other animals when they encounter the most dangerous predator on earth, man.

# MV *DONA PAZ* AND REQUIEM SHARKS

By Stephen H. Foreman

It was 10:30 p.m., December 20, 1987, in the waters of Tablas Strait near Leyte Island, the Philippines, a maritime nation with an encyclopedic record of disasters at sea stretching far back into recorded history. The moon was in a new phase, which meant there was zero ambient illumination. The night sky was pitch black, a preferred situation for sharks that prefer to hunt at night. The oceanic whitetip and the tiger shark in particular inhabited these waters: warm, temperate, tropic (68 to 82 degrees Fahrenheit), with extreme depths of 2,000 feet or more. The oceanic whitetip has been observed diving to that depth but usually spends its life cruising the top 490 feet of open ocean. At that exact time, there were two ships on a collision course, the MT *Vector*, an oil tanker with a crew of sixty-six, and the MV *Dona Paz*, a ferry, carrying 4,386 passengers.

The passengers on the *Dona Paz* departed Tacloban on Leyte Island and were heading to Manila to celebrate

Christmas with families and friends. Most of them were in a festive mood as they crowded onto the ferry in anticipation of enjoying the holiday season with loved ones. They were resting, many sleeping, some four to a cot due to overcrowding, when without warning the MT *Vector* plowed into the port side of the MV *Dona Paz*. Both ships burst into flame within minutes of the collision. Passengers were either burned to death or drowned. Corpses found bobbing and floating in the aftermath were charred over most of their bodies, many beyond recognition. Only 108 dead bodies were recovered from the sea, a tiny fraction of those who had been aboard both ships. Only twenty-six passengers lived to tell the tale, two from the sixty-six man crew of the Vector, twenty-four out of 4,368 from the Dona Paz. There was no way of knowing exactly who was on board because the overage was due to travelers boarding illegally without tickets under cover of darkness or the many who bought tickets once on deck. None of the names of these people would have showed on the manifest, such as it was. The nameless, charred, and mutilated corpses washed ashore on nearby islands where the locals respectfully buried them according to their native religious rituals.

The tanker MT *Vector* was fully loaded with a cargo of 8,800 US barrels of oil, gasoline, and kerosene. The *Dona Paz* was lethally overloaded with passengers. It was an accident that never should have happened. It was not an act of God, the weather, or a mechanical malfunction. Lawlessness, cutting corners, carelessness, and greed were the culprits—the exact same reasons why so many of our man-made tragedies occur. Journalists labeled it the Asian *Titanic*, the worst maritime disaster in the history of the world. With the *Titanic*, 1,517 passengers out of 2,223 were lost in freezing water of the North Atlantic. The *Dona Paz* lost 4,360 souls in water that was ablaze on its surface with 280,000 US gallons of disgorged oil. It literally was a fiery sea.

Both ships were travelling slowly: the *Dona Paz* at 26 km/hour; the *Vector* at 8km/hour. They were surrounded by thirty-seven square miles of wide-open sea, more than enough elbow room, so to speak, to avoid a collision had it not been for human negligence. Both crews appeared to have been celebrating the Christmas season early. The crew of the ferry was seen drinking beer in the galley. The *Dona Paz* carried no radio. Her lifejackets were locked away. Her crew was minimally trained for such an emergency.

Seconds after the *Vector* exploded in flames, fire quickly leapt to the *Dona Paz* and enveloped her, as well. Both ships were blazing while the crew of the *Dona Paz* ran around as panicked as their passengers.

Still, investigations after the event lay the blame with the *Vector*. She had no license. Her lookout was not a qualified Master, and she was deemed unseaworthy before setting out. Both ships sank in 545 meters of water, 1,788 feet, prime depth for pelagic (deep-water) sharks. The *Dona Paz* went down in two hours, the *Vector* in four. No ships came to the rescue for eight hours. With the exception of the victims, no human being even knew the collision had occurred, although it stands to reason that the resident sharks did.

Four thousand, four hundred and twenty-six people—what happened to them?

These were heavily shark-infested waters with one of the largest and best-studied families known as requiem sharks, a family that embraces somewhere in the neighborhood of sixty separate species. Depending on the specific species, requiem sharks grow from four and a half to twenty-five feet, 850 to 1,900 pounds, and they eat everything. These include the tiger sharks and oceanic whitetips. Tiger sharks are known as swimming garbage cans. Coal, tin cans, bones, books, clothing, wine bottles, chicken coops, drums, unexploded munitions, rubber tires—all these items have been found in their bellies. Carrion is also a delectable morsel—human carrion, as well as animal. Human remains have been found in their bellies, and human carrion is what was floating in the water the night of the deadly collision.

Tiger sharks are only second to the great whites in attacks on human beings, or so most experts say. Actually, the oceanic whitetip probably accounts for more. Whenever an airplane or a ship goes down in open ocean waters, the whitetip is there, its jaws eager to clamp shut and devastate the victim's flesh with distinctive sawing motions that enable it to tear off hunks of meat and swallow them whole. If the tiger shark is the swimming garbage can, the whitetip is the trash compactor or food processor. These were the species that attacked the survivors of the troop ship USS *Indianapolis* after it was torpedoed by the Japanese submarine in World War II, probably the most infamous shark encounter in history.

To be fair to the deadly bull shark, another member of the requiem family said to account for more attacks on human beings than others, it may be true that the tiger and oceanic whitetip are a particularly devastating duo, but the bull shark inhabits shorelines and is able to swim inland in freshwater, thus gaining greater opportunity and access to hominid prey. Bull sharks, or also known as the Ganges shark, are feared in the Ganges River in India, for example, when Hindu pilgrims enter the sacred waters. Bull sharks attack in murky water, and the Ganges (or inshore beaches) are perfect habitats for them to make their strike.

Regardless, it is a certainty that either tigers or oceanic whitetips fed on fresh killed and charred carrion the night of the *Dona Paz* tragic disaster. One hundred and eight bodies were recovered, most showing shark bites. What happened to them? They drowned. They burned. And that night one, possibly two deadly species of requiem sharks sank their teeth into human barbeque. At least, as corpses are already dead, these unfortunate folks were spared the horror of an attack.

Requiem. What a strange name for a shark! A family of deadly predators named after a holy, somber, and heartfelt celebration for the repose of the souls of the dead. In reality the word derives from the French "rest" meaning death or "reschignier" meaning to grimace while baring teeth. Both definitions seem superbly appropriate. In their own inimical manner, the sharks celebrated the dead.

Yes, it's true that most sharks are not dangerous to human beings, even the sand tiger shark, no relation to the tiger shark but closely related to the great white. Yet, the sand tiger shark has a deadly proclivity, as well—intrauterine cannibalism or adelphophagy, which literally means "eating one's brother." In this case, the most developed embryo will feed on every one of its siblings. That's why, at its actual birth, it sports a fully formed set of teeth. Whichever is born first is the one that will survive.

Prevailing wisdom has it that you are more likely to be hit by lightning than attacked by a shark. Nonetheless, many sharks do attack, seemingly unprovoked, only who knows what the shark is "thinking" when it finds a strange creature frolicking in the surf or swimming in open ocean? This is their territory, and you have invaded their underwater province, so when such an invasion takes place, an attack is possible depending upon the time of day or the hunger level of the predator. Think of it this way. It is the middle of the night. You and your family are sound asleep on the second floor of your home when suddenly you awaken to a crash of glass and discover a burglar has broken into your house. What would be your response? You would grab your shotgun and God help the thief who stalks upstairs and jiggles the knob of your bedroom door! Given a parallel circumstance, expect the shark to strike when its home is invaded.

And, yet, here is an interesting item: statistics from 1987, the year of the *Dona Paz/Vector* disaster, report that, globally, there were eighty people bitten by sharks, but there were 1,587 cases of humans biting humans (and that was recorded in New York alone). Human bites are the third most common bites seen in emergency rooms after dogs and cats. Moreover, the risk of infection and disease from human bites is great and can even lead to the destruction of joints—in other words, human bites might well be the first step in long-term suffering. Is there any good news here? If one is fatally attacked by a shark, there is no long-term suffering, no staph infections, no hepatitis C and B, no herpes, HIV, or rabies—perhaps you'll experience the horror of an imminent attack, feel an instant of pain, and then certain oblivion. One looks for the bright side wherever one can find it.

Here is still one more interesting item: the skin of a shark, when magnified, is made of tiny but exceptionally sharp teeth, denticles, replicas of those in the mouth, millions of very tough, minute scales that form a mesh of sandpaper-like protein, the exoskeleton and armor of the bearer. Even the skin of the shark can cause substantial damage.

Because sharks play such a common role in film, literature, and television documentaries, it seems as if we know a lot about them. Except we don't. Sharks are mysterious creatures. They lead secret lives. Given the advent of tagging, scientists can track individuals for months; however, sharks live on for decades after the tag is no longer functional. Therefore, we don't know many of the crucial details of their daily existence, for instance, their migration and mating patterns.

Also, finding them is not easy. Their travels that can take them immense distances are mostly hidden beneath the water that separates our world from theirs. The appearance and disappearance of this fascinating beast are usually unexpected. Perhaps that is why our fear is born.

We don't know where they are. We don't know if they are down there looking up at us, or down there at all, only to erupt from the water when we least expect it, shredding the peace of our dreams and forever altering our lives. That gaping jaw of this apex predator, that killing machine with its mighty phalanx of merciless, dagger-like teeth poised to snap at 1.8 tons of force (twenty times greater than the force of a human bite), must rank at the top of the list of the most frightening sights ever experienced by a human being.

And if, suddenly, this miscreation, this fiend bred in the black depths of Hell, appears next to you? Its coal-black and malevolent eyes gleaming in a refraction of light through the ocean's surface, its mouth agape revealing jagged, angular, misshapen, horrendous teeth in what looks in an instant to be a frenzied smile. A sudden surge toward your midsection as you flail and punch . . . you feel searing pain and the water clouds with bubbles from your exertions but also from your own blood. In that moment, do you think of man's dominion over wild creatures, and all we've proved of our species since Neanderthal days through our growing and expanding intelligence and accumulation of knowledge? It's all meaningless in the face of this brutal attack, and you then understand that the shark only wants to survive and you're an invader and also food. This is not your world, you've intruded into a foreign realm—Inner Space—and you've been eliminated, your life has ended by tooth and jaw. That fits the very definition of unspeakable horror.

# SS CAPE SAN JUAN

*Though considering that more than 1,400 men were on board the* Cape San Juan, *the casualty toll was not overwhelming after the ship was torpedoed off the Fiji Islands in 1943; initially, 117 men died from the explosion and its aftermath, although many faced their end at the gruesome jaws of sharks. Rescuers who arrived hours after the initial explosion found men in the sea in rafts or clinging to life jackets, blinded by fuel oil, and defenseless against sharks.*

As horrific, tragic, and terrifying to contemplate as the sinking of the USS *Indianapolis* was in 1945 in World War II, before that came the sinking of the troop transport ship SS *Cape San Juan* in 1943 in the South Pacific. There are many parallels between the two disasters. A submarine (Imperial Japanese Submarine *1-21*, led by Commander Hiroshi Inada) torpedoed the *Cape San Juan* on Thursday, November 11, 1943—now known as Veterans' Day in the US and Remembrance Day in Canada—as an Imperial Japanese Submarine did to the *Indianapolis* (Imperial Japanese Submarine *1-58*) two years later at the end of

June and beginning of July, and near the end of the war. The *Indianapolis* delivered components of the atomic bomb dropped on Hiroshima, as reported in this book's chapter "What *I-58* Caused." The *Cape San Juan* was about 300 miles off Fiji in the South Pacific, unescorted, heading to Townsville, Australia. Many of the 1,464 aboard (one newspaper report said 1,429) belonged to one of three units of the US Army Air Corps:

855th "All Negro" Engineers (Aviation) Battalion—811 officers and enlisted men;

1st Fighter Control Squadron—367 officers and enlisted men;

253nd Ordnance (Aviation) Company—162 officers and enlisted men.

While the sinkings were similar, the aftermaths were completely different.

A common consequence of both wrecks was fuel oil pouring onto the water's surface from the ships—it was a nuisance and a blight, coating the *Indianapolis* survivors and sickening them; but it blinded many of the *Cape San Juan* survivors, rendering them sightless and easy prey for marauding sharks. The sharks appeared quickly and seized on the *Cape San Juan* survivors immediately. The *Cape San Juan* had 1,464 on board (records show different numbers, but this number was common to several historical reports), and in what might be considered a maritime miracle, many survived and were rescued—meaning other records show that 117 died from the torpedo blasts or from sharks or drowning. One online sources tells a different story, reporting that fewer than 500 survived. However, it appears that the rescue ships and seaplane saved more than that. As many as 800 could have been lost. It's clear that more survived than after the sinking of the *Indianapolis*, after which fewer than 400 out of 1,100 on board survived the explosions of the torpedoes, the exposure of being adrift in the sea for days, the associated madness that set in among many—and then the sharks, thought to be oceanic whitetips. No less a marine-life authority than Jacques-Yves Cousteau advised that oceanic whitetips are especially dangerous because their curiosity makes them bold: they don't seem to possess a fear of man. Maybe they simply have no fear at all. Although other sharks seem to circle or ambush their prey, oceanic whitetips want to inspect their meals first and often bump or nudge the prey before attacking. This is a fearsome attribute that creates the impression of a robotic eating machine relentlessly in pursue of a kill. All the survivors of the *Cape San Juan* in the water could do was wait their turn, pray, or thrash in an attempt to frighten away the sharks. Their ship was sinking, they were adrift in the South Pacific, and they were surrounded by terror—from the sharks and from an enemy submarine likely still in the area. Would they be eaten by a shark, or shot through by a Japanese machine gun?

The *Cape San Juan* was on its second voyage when it was sighted by the Japanese submarine *I-21*. The ship's history was: It was delivered June 10, 1943, to the American-Hawaiian Steamship Co. and was almost

420 feet long, a conversion of a small dry-cargo freighter. It was slower than other Navy ships, and perhaps that made it a target for a submarine attack.

The configuration of the ship, according to Master Walter Mervyn Strong: "An additional deck house had been built on deck over troop exits thru 1, 2 and 3 hatches. This deck house was used as a washroom and was also used as a means of exit from the troop quarters in #1, 2 and 3 upper 'tween deck."

Strong reported that "the deck house covered a deck area extending fore and aft over the watertight bulkheads where they were fastened to the main deck. The purpose of this was to cover the troop exits that came from #1, 2 and 3 'tween decks with the one deck house. The deck house was watertight to the sea and weather, but no provision was made inside the deck house itself to separate the deck openings, which were close together."

Of the three full decks below the main deck, it is assumed they were full of cargo. The ship had six life-boats, four wooden rafts, and thirty-six Carley floats that held twenty, forty, and sixty people. Everyone on board had a life vest and lifesaving suits were available, though not needed because the location was the South Pacific.

The ship was under a blackout order, and under orders to zigzag. It was turning to starboard when the torpedo hit. Various eyewitness accounts described what followed:

*"Two water spouts seen at a distance of approximately 2,000 yards; relative bearing 120 degrees. Wake seen on water when 15 yards distant from the ship; very straight path; approximately 2 feet wide; went aft of vessel, missing stern by 20 yards. Wake was light greenish color;*

*water itself was deep dark green. Left slight white foam on surface."*

Another eyewitness claimed to see the wake of this torpedo some 300 feet out on the starboard quarter and claimed that it missed the ship's stern by only fifteen feet. A few seconds after the first torpedo passed aft of the vessel, several armed-guard lookouts saw two water spouts, described as from six to ten feet high and from two to three feet wide. They were described as egg-shaped by one witness; as fan-shaped by another; and as being "in the shape of a pine tree" by a third. Both spouts did not rise simultaneously. The second came up as the first settled. These were described by some as "narrower than whale spouts."

For the Japanese submarine, the offensive occurred from a relatively short distance. More commonly, this type of submarine was known to fire a spread of two to three torpedoes from about twice the range as the *Cape San Juan* attack—*I-21* fired from hundreds of yards away, not the thousands of yards of a more typical attack; the distance was desired if not required to avoid detection by the enemy, of course—with the intent to immobilize the target. Then the submarine would release another one or two torpedoes to finish off the target. A piston system similar to the one perfected on German submarines eliminated a release of compressed air (clearly the water spouts mentioned in the above passage) during submerged firings, which would otherwise betray their position. Perhaps the submarine experienced a system malfunction, or an error was made by the Torpedo Officer, or they simply weren't concerned about it, so confident that the ship would be sunk. The *Cape San Juan* almost certainly defended her-self—the submarine was within range of the ship's defensive guns—with the Navy Armed Guard returning fire immediately, judging the location of the submarine from the direction of the torpedo. One of the 20 mm guns jammed, records show. All guns fired intermittently for about ten minutes. Occasional shots were fired throughout the day to let the enemy know troops were still aboard and would defend themselves.

After the torpedoes struck, an eyewitness account reported: *"The ship shook and shuddered and the bow raised slightly, then settled and the vessel took on an immediate 10 degree to 15 degree list to starboard, and then settled to a 20 degree to 25 degree list within a few minutes."*

Reports were that a torpedo struck below the troops from the 855th. The hatch covers over the No. 2 hold were blown upward and then collapsed down into the hold, killing and injuring several men. An SOS was sent along with the message: *"torpedoed, ship sinking fast."* According to an account of the attack, inexplicably, the Officer in Charge of the radio, Lt. Harris, ordered the radio destroyed immediately afterward and abandoned ship, so no further signals were sent. Some effort was made to repair the equipment, but it appears unsuccessfully.

The ship's engine was ordered stopped, and the ship coasted and came to rest as she continued her turn to starboard. Mainly the enlisted Army personnel were evacuated about fifteen to twenty minutes after being hit, and those aboard expected further torpedoes hits. Just then, the seas became angry, surging to fifteen feet. Where the strike occurred at the No. 2 hold, the flooding began quickly, and the ship settled down by the bow with the starboard list increasing.

Officers of the 855th included Captain Herbert Edward Bass—Bass was a Lieutenant at the time of the attack, in charge of Company A—Captain Wholley, and 1st Lt. Mutchler and enlisted men including Sgt. Chester L. Rivers, First Sgt. Shelton, and Private Monroe Barkley. Bass and Barkley in particular showed remarkable courage by diving into the dark, oily, flooded hold to tie ropes to an injured man (Theodore Harris) so he could be pulled out. Doctors on board, 1st Lt. John G. Schurts, 1st Lt. James V. Davis, and 1st Lt. Leo S. Wool, set up a dressing station on hatch No. 4 to care for the injured men. Major Floyd C. Shinn, the senior passenger officer, was also singled out for his courageous actions.

It is unclear how many men perished immediately after the initial torpedo explosion, subsequent flooding, and the collapse of the hatch structure. Bass estimated twenty at the time, but most sources now agree on sixteen. Some men drowned while they abandoned ship by jumping overboard in full combat gear, and others were lost in the water near the ship when large wooden rafts were released over them. Understandably, most men in the water and even in the rafts tried to get to the few life boats, which soon became heavily overloaded. The No. 4 motor boat was swamped and lost when too many men (estimated about sixty-five) attempted to board it, and another raft came very close to capsizing. Some rafts drifted away before they could be manned.

Many of the men initially ended up with no aid but a life vest. These were a mix of cork and kapok. Years later, survivors Chester Driest and James Reed would laugh at how some of the life vests were stenciled "For Inland Waterways Only," including Reed's. Wind and wave action quickly dispersed the men in the water to the south, and the oil caused severe eye irritation and even temporary blindness in some cases. (Keep in mind, *Cape San Juan* was carrying six lifeboats, four large rafts, and thirty-six smaller rafts.) Soon, all lifeboats and rafts were launched, with the exception of those reserved for the 200 or more still aboard the ship.

About two hours later, reports indicate that a dull *thud* was heard throughout the ship, possibly a torpedo that struck the ship but didn't explode on impact; the fact that an explosion was heard moments later would explain the unexploded torpedo glancing off the ship's hull and exploding underwater—a bit of fortune amid the ensuing turmoil. Soon, a patrol bomber circled the ship and returned about two hours later signaling help would be coming.

That help turned out to be Navy Liberty ship *Edwin T. Meredith*, commanded by Master Murdock D. MacRae, whose initiative and courageous decision to enter the waters, where clearly an enemy submarine still prowled, saved countless troops—either 438 or 448 survivors, depending on the source. MacRae may have faced a reprimand or disciplinary action later for making this decision, though history doesn't conclusively show that, and records indicate that his Naval career continued. (He was later Master of the *Robert J. Walker*, sunk by a Japanese submarine in December 1944.) The Navy didn't have many ships available for the rescue of *Cape San Juan* survivors, as they were preparing for an invasion of the Gilberts, where the terrible bloody battle of Tarawa and the airstrip on Betio Island took place in November 1943. The survivors of the *Cape San Juan* were saved, you might say, by *Better Homes & Gardens*, as the *Meredith* was named for Edwin Thomas Meredith, the prominent Iowan who started *Better Homes & Gardens* magazine and built a publishing company, specializing in magazines.

MacRae piloted among the troops in the water and they grasped cargo nets lowered down the sides and *Meredith* soldiers and sailors aboard the *Meredith* assisted them to safety. Again, reports say 448 were ultimately saved in this manner. Sharks were everywhere, it seemed, and the troops in the rafts, many blinded by the fuel oil on the water's surface and now coating them, were being bitten or taken down by the sharks. An article from 1944 in the *Times* (San Mateo, California) had a headline "Man Eaters Pulled Living off Life Rafts" and "Survivors of Transport Sinking Tell Scenes of Horror" before documenting the horror faced by the men in the water. "I saw sharks grab two . . . from the *Cape San Juan* who were hanging on to the life rafts, and bit off half their bodies," said *Meredith* Second Engineer John Lopiparo. "There were a few bubbles and [they] went down. They couldn't even put up a battle because they had nothing to fight with. The men we rescued were sick and almost blind from fuel oil on the water." Sharks are known to have acute senses, and it could be that the fuel oil attracted them. In the book *Shark Attacks: Inside the Mind of the Ocean's Most Terrifying Predator*, author Gordon Grace states that a shark has a massive olfactory lobe in the brain, and it moves its head side to side as it swims to pick up scent in the water. According to the book, "[t]he oceanic whitetip, notorious for its habit of showing up at shipwrecks, is even said to poke its head out of the water to sniff for airborne scents." Further, the book shares that a "set of fluid-filled canals wind through its head, vibrating in sympathy with the surrounding waters. These canals serve the same purpose as our ears." This is probably how sharks detect underwater sounds so well, aided also by their lateral lines, which many fish have and use to detect prey or danger. The sharks looking for meals were fully alert to splashing and movements of distress and panic, and scents such as oil and blood. A report titled "Shark Behavior Still Cloaked in Mystery" in the *San Antonio Express News* in 1966 says, "A shark can sniff out an ounce of blood in millions of gallons of water, or detect a scent a quarter-mile from its source." Scientists further documented that sharks have electrosensitive pores in their snout, which helps them identify prey through electroreception—the electrical impulses are detected and sent along the lateral lines, most likely. This is an acute sense that most sharks have and is a tremendous advantage to feeding in salt water, which is a good conductor of electrical currents.

Many of the survivors were standing waist deep in the rafts, as the rafts called Carley Floats had dropdown bottoms held by latticed, netting sides, which made the men obvious and tempting targets for sharks. Imagine the scenario: In raging seas, the rafts were low to the water, and men clung to each other or the sides of the rafts while standing up, kicking and screaming, inadvertently baiting the sharks into a feeding frenzy. When bumped by a shark, a man would begin to bleed from the coarse skin of the shark causing an abrasion. This scent would intensify the frenzy, no doubt. Many men, when they leaned onto the floats, couldn't see to evade a shark slashing at them. They were blind from the fuel oil in their eyes and therefore powerless to see and fight off an attack, and the sharks ate well.

The crew from the *Meredith* worked furiously to save the *San Juan* survivors, diving into the shark-infested water and dragging men to safety. "The sharks began converging on the rafts. The *Cape San Juan* gun crew fired into the sharks, but they couldn't scare them away," the report in the *Times* said.

"You couldn't see the sharks in the semi-darkness until they were 25 yards from you. Time after time I heard soldiers scream as the sharks swept them off the rafts. Some times the sharks attacked survivors who were being hauled to the *Meredith* with life ropes," read the report in the *Times*.

Imagine the futility of being a rescuer on the *Meredith* and seeing and sensing the desperation of the soldiers and sailors in the water. All they could think was, "I need to engage, help, and save the survivors—my countrymen in wartime need me." MacRae asked for volunteers to jump into the lifeboats, and his men responded. Many were later commended for valor; after the war, five Merchant Mariners were awarded Meritorious Service Medals "for Conduct or Service of a Meritorious Nature," the commendation read. MacRae continued: "One soldier told me: 'I was sitting on the edge of a raft talking to my buddy in the darkness. I looked away for a moment, and when I turned back, he wasn't there any more. A shark had got him.'"

MacRae said, "I wish to state my highest recommendation for the Merchant Marine crew of this vessel, the S.S. *Edwin T. Meredith,* also the Navy Armed Guard crew and Army Personnel passengers aboard in their assistance in rescuing survivors from the water and in many instances parting with nearly all their clothes." The clothes of the survivors were saturated with fuel oil, so sailors on the *Meredith* gave them clean clothes. The rescue ship stayed nearby until dark but left before the Japanese submarine that was assumed to be in the area could return under the cover of darkness. Hundreds of men remained on board the sinking ship, though the ship the *McCalla* completed the rescue the next morning.

The Chief Mate of the *Meredith* said, "Captain MacRae made one of the hardest decisions of his life when he decided to go to the rescue of the *San Juan*. He assumed that the submarine was still in the vicinity, and the decision he had to make was whether to attempt rescuing those men, and at the same time jeopardizing his own ship and the men under his command. We are all glad he decided that way he did." MacRae was quoted in a newspaper report: "We stayed with [*Cape San Juan*] until 8 o'clock that night. It was getting pretty dark. I figured the sub was still in the vicinity and everybody agreed I'd be crazy if I stayed there overnight."

MacRae said he hated to leave, as hundreds of men were still stranded with the crippled ship. The *Meredith* trained her guns on the *Cape San Juan* and opened fire at the waterline, in an attempt to scuttle the ship. It did not sink. She finally went down the following day, after the remaining crew was rescued by the *McCalla* and *Dempsey*.

The *Meredith* and its full load arrived in Noumea, New Caledonia, off Australia. A Naval Martin Mariner seaplane, allowed by Navy authorities to continue on its rescue mission to reach the *Cape San Juan* after initially being ordered to abort due to the bad weather, picked up forty-eight survivors. This was a case of a military supply "flying boat" coming to the aid of the crew of a supply "dry-cargo freighter" or troopship—brothers not in arms, but in supply. This was the war called by Roosevelt's Secretary of the Navy Frank Knox "a war of supply." Particularly in the vast Pacific, supply transport was key to the war effort, and America's readiness and ability to mobilize supplies undoubtedly played a tremendous role in its triumph. A destroyer picked up the rest the next day and the foundering *Cape San Juan* was shelled by the rescue ship and scuttled, as the heavy seas helped bring the ship down.

*I-21* was never sighted again and made its final report on November 27, 1943, off the Gilbert Islands (also known as Tungaru and now making up Kiribati). History shows that a Japanese Type B submarine was torpedoed and sunk off Tarawa on November 29, 1943. That was thought to be *I-21*, meeting its watery fate. Perhaps the oceanic whitetips followed the submarine down, hunting for human flesh?

# MURDER BY SHARK AT CHERIBON, 1945

By Stephen H. Forman

The atrocities committed by the German Gestapo and SS troops during World War II have become the stuff of hellish legends—the death camps, sadistic medical experiments, summary executions, firing squads, the gas chambers—all too horribly true.

Then there were the dreaded *Kempeitai*, Japan's secret police, the equivalent of the fearsome German Gestapo, brutal agents of the state who ruled over the occupied territories utilizing the methods of nightmares to maintain control over millions of innocent people. The legend goes, it was modeled after the French *gendarmes*, though it seemed to follow a philosophy of "by any means necessary." This applied in particular to the maintenance of loyalty during the war to the Emperor Tojo, who was a prior commander of the group in Manchuria. (The end of the war for Japan, much to the relief of humanity, brought an end to the *Kempeitai*, as well.)

The ghastly truth is that the Japanese and the Gestapo were equally murderous, yet, arguably, the Japanese were even more ingenious in their methods of torturing and executing the enemy—military and civilian—whoever the enemy happened to be on any given day. After the Japanese defeat in 1945, the *Kempeitai* destroyed thousands of documents, so the true extent of their atrocities may never be known. But what we do know is bad enough.

In 1945, Japanese troops attacked and defeated Java. The punishments inflicted on both captured soldiers and innocent civilians were so heinous as to defy comprehension. The punishments referred to were not spur-of-the-moment or heat of battle actions like bayoneting the enemy. These punishments were the premeditated products of minds as inhumane and bestial as any in the history of warfare. These actions demonstrated the human equivalent of our perception of sharks—pure, intense, soulless evil. It was as if Satan commanded the Japanese soldiers, crawled into their minds, and switched off any mechanism that empowered empathy and kindness and compassion, leaving ruthlessness and pitilessness in place.

Psychiatrists, criminologists, and forensic psychologists would call this collective behavior psychopathic. In 2013, the *Diagnostic and Statistical Manual of Mental Disorders* (the reference book called the *DSMV*) placed this condition under Antisocial Personality Disorders. The same cold, unfeeling behavior of the Japanese—totally devoid of remorse or guilt and a total disregard for the value of human life—is attributed to the worst and most horrific serial killers in human history in the US, living demons such as Ted Bundy, John Wayne Gacy, Dennis Rader the "Bind, Torture, Kill" or BTK Killer, Son of Sam David Berkowitz, Richard Ramirez or California's "Night

Stalker," and Jeffrey Dahmer. The criminal profile of these monstrous humans probably would fall into what's called criminal enterprise homicide or sexual homicide, whereas the Japanese military were likely in the category called the cause homicide class or mission-oriented homicidal types.

"And no one knows better than those who kill for policy, clandestinely or openly, as do governments of the world, which kill in the name of god and country or for whatever reason they deem appropriate . . . . I don't need to hear all of society's rationalizations. I've heard them all before and the fact remains that what is, is," Richard Ramirez said at his trial in the mid- and late-1980s. He described himself as evil and said he worshipped Satan. He continued: "I am beyond your experience. I am beyond good and evil, legions of the night—night breed—repeat not the errors of the Night Stalker and show no mercy." He added in another interview, "Serial killers do on a small scale what governments do on a large one." Perhaps this was an allusion to the Japanese or Nazis in World War II? "Killing is killing, whether for duty, profit, or fun," Ramirez continued. He remained in prison and on death row in California, though he never suffered the handiwork of a state-sponsored executioner: he died in 2013 from complications of B-cell lymphoma.

In the aforementioned serial-killer cases, one murderer acted on depraved or fantasy impulses. However, mass murder and suicide was the masterwork of Reverend Jim Jones and the People's Temple in Jonestown, Republic of Guyana, when 917 people died in 1978. Group murder is also a trait of human nature, the same as nature and animals.

The book *An End to Murder: A Criminologist's View of Violence Throughout History* by father and son Colin and Damon Wilson begins with an introduction by Damon: "There is something essentially wrong with the human race. And, ironically, it is in the light of our astonishing achievements that this wrongness is so clearly visible. . . . Creatively and intellectually there is no other species that has ever come close to equalling us." Tragically, history shows that the creative human mind devises horrific methods of killing other human beings. The author continues: "We are the only species on the planet whose ingrained habit of conflict constitutes the chief threat to our own survival: in Darwinian terms we are an enigma—a species so successful that we threaten our own existence." Wilson the son (writing in his native United Kingdom style, where the book was first published) provides an example of an appalling wartime killing in World War II:

> A young soldier was captured by the enemy. They discovered that he was an excellent pianist, so they sat him at the piano and told him to play. He was also told that the moment that he stopped playing, he would be taken outside and shot. The young man played continuously for over twenty-two hours, until his arms and fingers were in agony. Eventually he collapsed in tears, unable to play another note. His captors heartily congratulated him for such a Herculean effort. Then they took him outside and shot him.

Examine your emotional reaction at this moment. How much do you empathise with him? With his fear, pain, and his final despair when he realized that his captors' laughter and slaps on the back did not mean that they were doing to spare him. How do you feel about the men who tortured and murdered the young soldier? Can there ever be any justification for such heartless cruelty?

Now consider the following additional facts: the young man was a member of the Waffen SS—the Nazi Party's elite shock troops, who ruthlessly carried out some of the worst atrocities of the war. The Russians who tortured and killed him had just fought their way across hundreds of miles of scorched earth, and knew of compatriots by the thousand—non-combatant men, women and children—who had been murdered by the retreating Nazis.

Human nature can be shocking, vengeful, regrettable, puzzling, horrific. We the human race have a particular skill at determining tools or instruments of destruction, and rather than pulling a trigger or thrusting a lance or dagger ourselves, we outsource killing to bombs, poisons—or sharks.

If you were in Java during World War II you would have witnessed insane cruelty without a shred of remorse. A young Dutch girl, seventeen years old, a survivor as a prisoner of war, wrote in her memoir, "How can you dream when you are locked up in a dirty, over-crowded prison, when you are lying on a filthy mattress full of bugs? How can you dream when your stomach cries for food? How can you dream without music?" And she was fortunate because she escaped the worst. Women in Java captured by the Japanese were often raped multiple times before being disemboweled or beheaded. If they were lucky they were sent to brothels as "comfort women."

Behold what follows:

Put yourself in an occupied village surrounded by an enemy that hated you, an enemy you knew lacked the conscience of a savage beast. They seethed with their own delicious hatred.

It is July, 1945, and you are one of 150 civilian residents, primarily Dutch nationals, left in North Java. Japanese soldiers have defeated a force of Australian and British troops and have now occupied the territory. They have embedded themselves in your homes, your schools, your factories, your places of business. All adult males and teenaged boys, if they had not been already executed on the spot, have been rounded up and shipped to internment camps. The enemy soldiers have bayoneted your neighbors, shot your relatives, tied the thumbs of your friends together and then tethered them to an automobile or truck that pulled them around in a circle until they dropped. You know the enemy despises you, and they are everywhere. You know that you could be the next to die, and you can only hope—pray—death will come quickly.

One morning, for no discernible reason, you might imagine, everyone who's left in your village—you, your family, your neighbors, women, children, old men—are rounded up at gunpoint. Enemy soldiers are screaming at you in an incomprehensible language. People do not obey because they do not understand what they are being told. The person next to you is summarily shot in the head because she did not obey a command she didn't understand. A person a few feet away is casually bayoneted because the soldier decided he wanted to. Reason has disappeared from the face of the earth. Now, you are marched to the port of Cheribon. You don't know why. You are scared and confused. *What are they going to do to us? What is happening?* The only thing you know is that whatever it is will be horrible. That is the modus operandi of your captors. How horrible you do not want to think. You erase all thoughts and allow your mind to be inert. You empty your mind.

A Japanese submarine is moored at the dock. Uniformed guards are yelling. You still don't understand what they are shouting, but when they indicate with their rifles, you know. Board the sub. Stand on the deck. Ship's guns fore and aft are aimed at you. There is nowhere to go, no escape. You know—you accept—any minute now, you are going to die, but why did the soldiers march you onto the deck of a submarine? Submarines hunt enemy warships, but you are not a battleship, not a weapon, just a simple human being. A human being who wants to live with the energy of every cell in your body—but you know they will not allow that. Holy Mary, Mother of God, they will not allow that. Peace to His people on Earth, they will not allow that. Now and forever at the hour of our death, and it is time. You follow the queue onto the submarine. You think how odd it is to be walking atop a boat that spends most of its time underwater. Lurking in the deep, waiting to strike unseen, unleashing terror and death, like a shark.

Powerful engines fire up. The sleek boat leaves the dock and heads for the open sea. *Where are they taking us? If they're going to shoot us and toss our dead bodies overboard, why don't they just do it?* You are five miles out, maybe a bit more. Ocean. Vast. Deep. Then it happens. The submarine begins to dive. And you know, yes, you know, but still you cannot believe what you know is happening to you right now. Fear grips you in its iron vise. You can't think, but ice takes your heart. As the sub angles down under the water you are swept off the deck and into the sea. Terrible cries surround you. You scream, too. Why is this happening? You didn't do anything except have your home in what became occupied territory. There is no holding on. So what if you do? You will still drown.

Now you are in the water, treading desperately, trying to stay alive. Water invades your nose, your mouth. Your eyes burn from the salt. You are minutes, seconds away from certain death and, yet, your heart forces you to tread water. Your heart is not yet ready to die. You can't let go until it's taken from you. And then you see it—a white-tipped dorsal fin homing in on these people splashing, screaming, desperately trying not to die. Some have already drowned—the dead and the lucky. You wish to God you were one of them. You take a deep breath and sink below the surface. Please, you pray, please drown. You are bumped from behind. *Oh, dear God, please let me drown!* Oceanic whitetip sharks are attacking. You wheel around in the water, bubbles occluding your vision, lost in a swirl of panic . . .

You don't know this because you are no longer alive, but there was one survivor, minus an arm and a foot, of this massacre at Cheribon. He died shortly after being pulled from the water. One of the many executed by the Japanese *Kempeitai* (as mentioned earlier—military secret police corps, similar to the German *Gestapo*) in their brutal, inhuman fashion. "Our unit did things no human being should ever do," said one former member of the *Kempeitai* and Unit 731, which experimented with biological weapons in occupied China, in an interview decades after the war. "When I think of the people I killed so cruelly . . . I cannot help apologizing to them." All much too late, of course.

We hear of great whites and bull sharks being the primary cause of attacks on humans, but the truth is that the oceanic whitetip shark has killed more humans than any other shark in history. It's a numbers game, really—they've had more buffets served up to them. These are pelagic creatures, which means they only live in deep oceans, temperate waters of sixty-eight degrees, 150 to 450 feet below the surface. They rarely come near to shore, which is why they are not so well known. Their attacks take place in the open ocean on the "survivors" of ships that have sunk and planes that have been shot down. At first there is only one shark, and then there are many, snapping, biting, attacking in a frenzy. Throughout World War II, Japanese submarines and kamikazes made certain the whitetip and tiger sharks did not go hungry.

Oceanic whitetip sharks have a singular purpose: to eat to survive. Up to thirteen feet long and weighing as much as 400 pounds, with white on the tips of the dorsal and wide pectoral fins—the fin tips almost look luminescent underwater—they are horrific eating machines. They are similar to mako sharks; however, they stay only in the deep ocean. Their long fins, described as similar to airplane wings, allow the whitetip to cruise or glide through the open waters searching for meals—which are scarce in the open ocean, which makes these sharks particularly opportunistic about attacking. Their dorsal fins are rounded, perhaps from their constant prowling with the fin exposed above the water's surface. Scientists conjecture that the sharks can go weeks or months without food, so they seize every opportunity to feed, whether it's an injured fish or a shipwrecked boater treading water. They're known to be curious or inquisitive sharks, coming directly at divers to inspect them—possibly to determine if they've found a meal. They respond to sounds before they're close enough to detect smells and see motion on the surface. Though they feed on anything, three of their common food items are tunas, stingrays, and sea turtles, large-bodied animals; they also eat sea birds, again larger silhouettes seen from below against the surface light. They are formidable, horrifying, not least because of the mysterious environment in which they live—the vast expanse of open ocean, what scientists call inner space.

Encountering a whitetip shark means being trapped or stranded in the seemingly fathomless ocean—out in the open, nowhere to hide, completely exposed. The substance of nightmares.

# EPILOGUE

*More shark stories . . .*

# BLACKFISH AND HAMMERHEADS OF BARROUALLIE

By Stephen H. Foreman

*"Mongoose, mongoose, roll ovah.*
*Come out de watah deah. Yeh.*
*Dey come out de west, de watah deah. Yeh.*
*Come from de west. Yeh.*
*Mongoose, mongoose, roll ovah.*
*Come out de watah"*
—Chant of West Indian blackfishermen

Zekiel Millington sat in the stern of his boat, his forearm guiding the tiller, the hand of the same arm on the throttle of the outboard motor. Aaron Moses, the gunner, stood on the bow singing out to the whales below. Words with a West Indian lilt tumbled from Millington's mouth like a windchime made of bamboo.

"I see de moon when she high and big ovah de hahbor. De watah calm and de fish comin'. Not so when de moon be low on de watah. Watah rough, too rough. Fish not comin'."

But, the water was "not rough today," he said, and he was certain the "fish be comin'."

Millington, with the emphasis on the last syllable, was a tall, powerful man with cocoa-colored skin—tropical cocoa, the kind sold in open-air markets, thick sticks of the deepest brown—long arms and large, expressive hands that swiveled loosely from his wrists when he talked. They reminded me of first-baseman's mitts. Millington had the force and presence of an actor who is slightly mad and always unpredictable, as if the tropical sun had melted that small portion of his brain that dealt with impulse control. He wore a battered straw hat that might once have been a planter's. Like the rest of his crew, his shorts had been cut from old trousers. His yellow Lacoste shirt, probably a yachtsman's castoff, was so ragged you could see the gray nap of hair on his chest. The glare of the morning sun on the sea was already so intense I had to put on sunglasses, yet Millington's eyes were wide open. He barely blinked as he stared at the horizon line looking for a sign. I sat on the gunwale opposite him and tried to stay out of the way.

We left the bay at Barrouallie and headed west into the channel between the islands of St. Vincent and Bequia in the Grenadines. This channel as well as the one north between St. Vincent and St. Lucia is one of the waterways taken by the blackfish on its spring migration. The blackfish, which averages twenty to twenty-three feet in length and reaches three tons, is what the descendants of ship-wrecked slaves and Caribe Indians, like Zekiel Millington, call the pilot whale. Herring fishermen dubbed it the pilot whale because it led them to huge schools of the sleek, silver fish. Its cousin, a slightly smaller and much more aggressive version similar to the killer whale in temperament and related to it, as is the blackfish, they call the mongoose. Like bats, they echolocate their prey, and they are known to hunt in packs like wolves. The whalers sing it, cajole it, pray it, and curse it to the surface because the blackfish is their livelihood. These waters also hold the great humpback that grows up to fifty feet, and the massive sperm whale that reaches sixty. When Millington was younger and dreamed of having his own boat, the blackfish were the nuggets, the humpback and sperm the mother lode. But he is older now, and he wonders whether he still wants to go against a creature that size.

"Mongoose, mongoose, me comin'. She down deah. Come up."

Millington and his two-man crew are the last of the whale hunters on St. Vincent and among the last in this area of the world. The oldest, Veron, was sixty-nine; the youngest, Aaron Moses, was sixty; Millington himself was nearly sixty-five. There were two other crews still plying these waters: one off St. Lucia, one off Bequia, and these men were as old as Millington's. Athneal Oliviere of Bequia, acknowledged as the premier whaler and harpooner in the islands, was sixty-eight and had already declared this to be his last season. It was becoming dangerous, he said, because it was harder for him to put away the harpoon. The young men of their villages are interested in easier ways to make a living; and, so, when the old men die, there will be no one to take their place. This will be the last season for Millington and his crew as well, and they will hunt in much the same way his predecessors have for hundreds of years. There are two differences: a hand-made harpoon gun mounted on an iron quadruped on the bow, and an outboard motor. But, most of the time the twenty-three-foot wooden boat is under sail. She has no instrumentation whatsoever, not even a compass, and she has no cabin and no canopy, so the men are totally exposed to the sea and sky. She also leaks. Her sails are ragged. Her paint is chipped away. Everything is either salvaged or handmade, yet *Faith* (which is what Millington christened her) skips and glides through the water agile as a turtle. Her harpoons are wrought from rusted automobile springs. Her gun is literally the butt, chamber, and trigger housing of a cheap shotgun with the barrel removed, which has been fitted to a pipe banded to a thick slab of wood—the idea of a St. Lucian fisherman who had served in the US army during the Korean War. Once the gun is fired, the rest of the harpooning must be done by hand. The harpoon itself weighs sixty pounds. Regardless of your politics, this is a remarkable feat when you think of a man with a harpoon poised to strike balanced on the bow of a small boat as it plows through the water after a powerful mammal nearly the same size as the boat, or, in the case of the humpback or sperm whale, three times its size. But, it is even more remarkable when you consider the age of this man.

It was March, 1989. Millington had just agreed to take me to sea with him the day before, but it was a journey that really began nine years prior to that when a friend of mine on the island first told me about the blackfish crew. My friend thought I'd probably be interested in them, and she was right. I was committed to leaving St. Vincent the next day, but the blackfish were running, my friend said, and I wanted to see what I could see before then. We'd been having a beer, but I cut it short because I wanted to go right that minute while there was still light. Barrouallie was all the way up island, and I'd have to flag a van to get there.

None of the vans on St. Vincent followed a fixed or published schedule, but they departed the Kingston Square with regularity and ran from dawn until just after dark. Either you boarded one at the square or waved it down somewhere on the road. It didn't matter where. A stop was wherever a customer happened to be standing. Each van was owned individually. I could never figure out, on an island so small and so poor, how there were so many men with so many vans, but the vehicles were all impeccably maintained and emblazoned with hand-painted lettering front and back that declared their names: *Have Faith, Wonder Not, Who To Blame, Moment Of Truth, Labour's Reward, Endurance, Free Mandela, Revelation*. Barrouallie was so far up-island, however, and so self-contained (the villagers tended not to come into the city except for market days) that it was difficult to find a van going there. I finally did—*Dignity* (not *The Dignity*, just *Dignity*)—and left the city of Kingston about two-thirty in the afternoon.

*Dignity* looped inland at first, but within minutes we were climbing the hills along the coast on roads so narrow that any oncoming vehicle had to pull to one side to allow the other to pass. Not that this slowed anyone down. *Dignity* hurtled forward fearlessly. She played guts ball. Her driver handled her well, however. I could see this guy driving the presidential limo, so I sat back and hung on.

*Dignity* continued her climb along the narrow road with the coast on the left often hundreds of feet down. We passed hamlets and clearings with names like Redemption Sharpe, Hog Hill, Questelles, and Bamboo Gutter, crossed the Camden Park and the Cane Wood rivers, reached a portion of road that was so steep we stayed in second gear for miles. This was plantation country; and, although most of the large holdings were now gone, I could still see fields and fields of coconut palms sloping to the sea as evenly spaced as pickets in a fence. A woman with a large reed basket balanced on her head plodded slowly up a hill. A man led a small donkey loaded with firewood up the hill beside her. The woman and the donkey walked with the same rhythm, the same sway, the same passive determination. The man carried a cutlass, an item as common in the islands as chopsticks in China, with the same blade design as those carried by the buccaneers hundreds of years before. We passed Anse Cayenne and Rilland Hill, Chauncey, Byahaut, and New Peniston on the Buccament River. The road dropped sharply to sea level, where we suddenly seemed to be hemmed in by bamboo thickets and then banana trees. It was torn and potholed, the result of hurricanes and poverty. When it opened up again, the Buccament Escarpment loomed before us like the Leviathan, like some enormous prehistoric monster. It faced inland with its head a triangular-shaped point of volcanic rock, its humped back covered with craggy plates, and a massive tail that ended in the sea. Its slopes were thickly matted with rainforest, luxuriously green, dense, and dripping wet, as if the beast had stretched out under the sun to dry. *Dignity* turned toward the sea once more and climbed a pass that cut through the beast's tail. A schooner had put into a black sand cove below us, and the sun was so bright the white sky weighed upon the water and blinded me. I shut my eyes. When I opened them again we were making the descent into Barrouallie.

This was a very poor place. Wooden houses, sadly in need of paint and repair, were situated all around the village square. *Dignity* stopped in front of a tiny food shop. It was about four o'clock. I got out and walked into the shop to ask directions. A woman with rum-colored skin stared sullenly at me from behind the counter. On the floor behind her was a corrugated box filled with small, torpedo-shaped loaves of bread and behind that a shelf stacked with tins of corned beef. I asked her where the fishermen would be putting in. Instead of answering me she pointed across the square toward the black, sandy bay where, apparently, all of the fishermen in the village kept their boats. I bought a loaf of bread and a bottle of Hairoun beer. As I crossed the square, I looked back over my shoulder toward the shop and saw the woman staring at me from the doorway.

I cut diagonally across the square, which a group of nearly naked children playing kickball shared with a herd of tethered goats and a mangy clutch of chickens, reached the beach, and walked out onto a battered concrete jetty to wait for the fishermen. Large nets had been spread out on the beach to dry. More chickens scavenged among them for bits of fish that had stuck to the green cords, small fish with names I had yet to learn—dodger, jackfish, sprat, and robin—baitfish used by the blackfishermen to lure the larger bonito and skipjack that they caught by trailing a line behind their boat. A large skull, bleached and weathered, sat by itself in the black sand. It must have been two feet across with eye sockets wide as cups. Its jaws were like a bear trap, different than a shark's but just as unnerving. It gaped up at the sky above the horizon like a ritual mask.

It was five o'clock. I had been waiting about an hour. The edge was just coming off the heat of the day. The sea was calm, pewter-colored in the distance, dark green as it was enclosed by the bay. There was no breeze at all. I looked for some sign of an incoming boat, but I couldn't see a thing. I had been told it was possible the fishermen wouldn't come in at all. Sometimes they stayed at sea for two or three days at a stretch. Since they had no radio communication system, there was no way of knowing whether they were pursuing fish or had gone to the bottom. Nobody knew until they saw them again. Or didn't. Some children were playing cricket on the beach with a slat from a crate, but no one else was walking about. It was dinnertime, and smoke from cook fires curled from the houses and small pits dug into the sandy backyards. For a minute more the sea was empty; and, then, there it was—a wooden boat at the very outer edge of the bay slicing through the water with a make on the jetty. Its sail was down, and I could hear the *put-put* of an old outboard as it pushed the boat along. The children playing cricket stopped their game and ran to the water's edge. Other children joined them, as did quite a few adults. They just stopped what they were doing and came to the beach. It was like a communal sigh of relief. The sun rose; the fishermen went out. The sun set; they came back. The boat came near enough to shore for everyone to see that today it was empty, so the villagers left the beach as quickly as a suburban crowd when a movie's over, without even a look back at the final credits.

The hull slid from the surf and skidded through the sand with a hiss. Even before it came to a stop half-in, half-out of the water, I had already decided I wanted to go to sea with them. There were six men in the boat, and it was obvious they were hunters—as all good fishermen, whether they are after a fifty-foot cetacean or a Dolly Varden trout, are hunters. They moved with grace and silence as they leapt from the boat and began unloading it. There was no rush, just method and purpose and complete attention. It was obvious, too, that this was something they had done hundreds, even thousands, of times. One shouldered the outboard, another handled coils of rope, but the one I watched was the gunman. The piece looked about as delicate as an anvil, but he folded an oiled cloth around each part as a jeweler would black velvet around a precious stone. All of the equipment was stored in a cinderblock shed at the back edge of the narrow beach. Not one of them paid any attention to me. They had to have seen me—they beached within ten yards of the jetty where I stood—but my presence was never acknowledged, not in a way that I could discern, anyway. While the others stored the equipment, one of them took a long, stout pole and punted the boat to its anchoring place in deeper water. I watched him jump into the water and begin to swim back to shore when I sensed something behind me. I say "sensed," but, in truth, I'm not sure what alerted me. I may have heard the padding of bare feet on concrete or the clinking of equipment, or maybe it was my personal radar signaling that the space around me had changed. I turned and saw the fishermen—all except for the one in the water—walking up the jetty toward me. They were grim-faced, and they moved quickly. I knew what I wanted, but did they? It suddenly felt very dangerous out there with nowhere to go except through them or into the sea. In an instant the distance between us would be closed, and I felt I had only that single instant to establish myself as a person they would either welcome or not. My mind raced. I had to bring it to a halt, and what it said was, "Be still." It said, "Don't *do*. Don't posture. Trust who you are. Simply meet them."

We met in the center of the jetty, and I have a vague memory of shaking hands. I remember them smiling. Their teeth were rotten, most of them missing, and it occurred to me that, perhaps, one of the reasons for their grim faces was that they hadn't wanted me to see how decrepit their mouths were. It amazed me that the mouth of the whale skull on the beach was in better shape than theirs. I do not remember what we first said or how we introduced ourselves to one another. It seems to me now that one instant we were on opposite ends of the jetty and the next they were showing me the whale gun with pride and telling me how it worked, and I was telling them about hunting deer and elk in snow-covered mountains. I hunted because I sought in myself what the animals always were, that they lived where I sometimes visited. I searched out this existence when the season came, when the game laws would let me. I needed to know I could take game because I did not feel complete without this knowledge.

The blackfishermen needed to take game because they would not be alive without it. I hunted to feel. They hunted to eat. But hunting was not our common bond. It was much simpler than that: we liked one another.

I doubt that we were together more than an hour that first day, but I asked if I could come back and go to sea with them. They said yes, no problem, just make sure I came when the blackfish were running. It was nine years before I was able to return; by that time, all the men of this original crew were dead.

How was I to know, when I first met the fishermen, that I was soon to start a new cycle in my life, one that would keep me from returning to St. Vincent for nearly a decade? I was emerging from a period where the impulse to act and the action itself were very nearly simultaneous. If I wanted to do something, I did it, and I had managed to survive. Good fortune kept me alive when good sense was in remission, but there were moments I would have considered myself more fortunate if it had been otherwise. What a mistake that would have been! This new cycle included a new marriage, a farm in the Catskills of New York, and a family. I fully intended to go back to Barrouallie the following year, only I was suddenly very busy nesting. Barrouallie assumed dream status—the "mongoose," wide around as a truck tire, and the men who read the water as easily as I could read a book. As my life became more settled, more bound to the earth, I would think of them not out of envy, but in wonder at the differences in our lives. When it was 90 degrees in the West Indies, it was 20 below in the fields surrounding my house. When the agonizingly beautiful melancholy of autumn settled over our valley, hurricanes tore across the island of St. Vincent. The mountains around us were old, worn, and rounded; thick rainforests make the mountains of St. Vincent appear lush and accessible. However, this is deceptive because their mountains are steep and young and, therefore, still reforming themselves. Soufrere, the largest of these mountains, is an active volcano that has already exploded twice this century pulverizing villages, torching the forests, and covering the natives with a pasty gray ash so thick it had to be scraped off. Soufrere dominates the island like Skull Mountain does in *King Kong*. And, yet, their climate is so temperate that if a seed merely drops to the ground it will sprout full and bear fruit, while my growing season in upstate New York is a scant couple of months.

The desire to go to sea with these men as they hunted the blackfish never abated. It was just bumped and tabled by more immediate realities. Life was different but certainly not dull. My wife and I had gone to Colombia and adopted a child, which turned into an adventure through the systems of men—courts, social workers, government bureaucracies—an infuriating and frustrating complex of legalities designed to keep you from the child you want. But a son was eventually the issue of all this. Then facing another winter in the Catskills where it took an hour to dress the boy for a fifteen-minute trip outside, I began to seriously consider the possibility of going to St. Vincent again. I called a friend who owned a marine supply house in Kingston, the capitol, to discuss it with him, and when he told me the fishermen were retiring and dying off, that only a few were left, and that their season was coming to an end, I knew I had to go back. My wife knew it, too. She had only one condition—that all of us go.

We left the Catskills early on a dreary morning, the first day of March. The sky was drab, deadeye drab. A veneer of ice from a freezing drizzle coated the trees, the road, everything. When you stepped on the grass it crackled. This was the time of year when the cold seemed as if it would last forever. Admittedly, the Catskills are not the Arctic Circle. There are colder places in the world than these mountains in winter. But our snows start about Thanksgiving, continue through April, and are known for an occasional roar right into May. They are deep and blinding. The thermometer plummets to 20 degrees below zero. Skin sticks to metal at that temperature. Again, admittedly, this is not 40 below, but I defy you to walk out on my front porch and tell me the difference. As far as I was concerned, it was the perfect time to head for the gentle weather of the Grenadines. The rainy season was another month off, and light winds floated in from the sea to keep the islands mild and pleasing. Slippery mangos and juicy grapefruit waited to be plucked off the trees, and my wife would sleep the sleep of angels without the goose-down quilt, which covered our bed twelve months of the year.

We flew directly to Barbados and then took a local puddle-jumper over the sea to St. Vincent. Old friends, descendants of the British who colonized the island, met us at the airport. After staggering through customs (an event that included a change of diaper, six milk crackers, a jar of strained spinach, and the frantic retrieval of Mutsy, the stuffed animal that had fallen off the stroller to the tarmac on our walk from the plane to the terminal), they drove us first to get a car then to the little house in a section named Rose Cottage where we would spend the month. We settled in and immediately drove to the open-air market on the other side of Kingston, the capital and largest city—actually, the only city on the island—a teeming brew of buildings, most of them one story, none more than two, narrow cobblestone streets, donkeys, vans, and old cars. There were sidewalks, too (depending on the block), but the citizens didn't seem compelled to use them, as there were just as many people walking in the street. Traffic was supposed to drive on the left, due to the British influence, though it didn't seem to matter to the drivers which side of the road they were on as long as they were pointed forward. Our friends had thoughtfully put in a basic supply of food and provisions for us, so our purpose in heading for the market was not to shop but to find a woman named Esther Irene. She operated a fish stall in the marketplace. According to inquiries placed by our friends, Esther Irene knew the whalers well and would, if she felt like it, facilitate meeting them. Finding her was the next step.

Only foot traffic was allowed in the space designated for the marketplace, so we parked the car, and, armed with the stroller, Mutsy, and diaper bag, we joined the swarm of islanders buying and selling their daily bread. The marketplace was a medley of rickety stalls and blankets spread on the ground displaying coarsely ground chocolate sticks, West Indian hot-pepper sauce, *dasheen*, *tanya*, herbs, beans, peppers, tomatoes, grapefruit, mangos, squash, bananas, and the morning's catch of fresh fish. I wondered if the voices of the people sounded as beautiful to one another as they did to me, that exquisite West Indian lilt that transforms even the most common string of words into an incantation. And nobody seemed to be in a hurry. They ambled from stall to blanket, dawdling over this and that, quibbling and sweet-talking, sometimes exchanging one parcel of goods for another, all on island time.

With my son spearheading the way in his stroller, we negotiated our way through the packed marketplace. Where was Esther Irene? We asked at each stall, each blanket. Everyone seemed to know who she was, but no one seemed to know where she was. A little girl tapped me on the arm. She was eight, maybe nine, shoeless in a raggedy dress with tight black braids like spider legs every which way all over her head.

"You lookin' for Miss Irene? Miss Irene's over there."

The little girl pointed off into the crowd, but I couldn't tell at who or what. I shrugged. She seemed exasperated. Without another word, she took the handles of the stroller and made her way into the thick of things. She carried herself royally, and I'm sure she thought herself very important, which, indeed, she was. We followed along behind her as she picked a path through all those people. She stopped at the blanket of a very old man who was displaying an assortment of wonderfully carved bamboo flutes. He put one to his lips and played a few notes. The tone was delicate and lovely, like a birdsong floating from deep within the forest.

He withdrew a new flute from a straw bag and handed it to my son, who promptly tried to stick the entire thing in his mouth sideways.

"You give him what you want," the little girl said, indicating the old man.

I took five BWI dollars from my pocket. The currency was pronounced "BeeWee," short for British West Indian. My wife said it wasn't enough. I counted out five more. The old man took them and thanked us with a wink of his eye and a trill of his flute.

My son put the end of the flute in his mouth and tried to blow through it as if it were a bugle. The little girl pressed on through the tumult of the marketplace. It was difficult to steer the stroller over such uneven ground, but her course was resolute, and my son bounced along happily sucking on his flute as if it were a peppermint stick. Soon we stopped in front of a rickety stall laden with mounds of fresh fish. Presiding over this with a machete in one hand and a voice that countenanced no nonsense was a chocolate-skinned woman no more than five feet tall, no less than two hundred pounds. Her dress was burnt orange, full but short, knee-length. Her hair was wrapped in a torn, red square of cloth. A tarp on the ground beneath her feet was littered with dozens of fish heads that she lopped off as easily as Madame Defarge. She hawked her catch like a sportscaster announcing the day's lineup.

"Miss Irene." The little girl tugged at her dress. "These folks want to meet you."

Esther Irene put down her machete and stretched out her hand. I took it. It was slimy with fish, but her grip was very strong. I told her our names and what I wanted.

"All dead but three," she said.

Would they take me out with them, I asked? She didn't know.

"I ain't dem," she said.

Would she introduce me? She would. I was to return in an hour when the market closed and drive her back to Barrouallie. She couldn't guarantee that the whalers would be there because if they harpooned one of the bigger whales, the giant might well tow them around for two or three days before they killed him. However, if they were there, she would certainly introduce me.

An hour later I was driving along the road to Barrouallie, the dangerously narrow coastal road I had traveled ten years before, the only road, potholed, cracked, in terrible disrepair. I had driven my wife and son back to Rose Cottage, set up the porta-crib, and kissed them good-bye. Most dads kissed their families good-bye to go to an office. I thought how fortunate I was to be a writer and so have an excuse for such an escapade! How many other kids had fathers with a day job on a twenty-three foot wooden whaler, or mothers who encouraged such a thing?

The road rose a final time then dropped steeply into Barrouallie. Nothing had changed. Goats still grazed on the rectangular green at its center; children still played there with a beat-up ball. The beach and sea lay just beyond. We parked next to the concrete jetty where I first met the whalers ten years before. It was late afternoon, but they hadn't put back in yet. Other fishermen had, however, and their families were busy stretching their nets out upon the black sand to dry. I felt a rush of déjà vu. A decade was as nothing. The sun rode low over the sea, and the water was just beginning to take on a glow from a sunset still an hour away. I took out my camera, but Esther Irene cautioned me against taking any pictures.

"They don't like none of that," she said.

So I put the camera under the front seat, locked the doors, and walked out onto the jetty to watch and wait. It was so quiet. The people on the beach barely made a sound. I must have blissed out for a few minutes because I was startled back to consciousness by the muffled sound of what was still clearly a boat's motor. A sharp whistle brought my head around. Esther Irene, with two fingers in her mouth and one hand pointing toward the horizon, was signaling me to look. The whalers were heading in from the sea. The people on the beach stopped what they were doing and looked, as well. They hung in a state of anticipation. Even the dogs stood still. It was obvious. We all wondered the same thing: Had they gotten one?

The sun was behind the boat, making it impossible to stare for very long; but, suddenly, one of the nearly naked little boys who had been squatting in the sand jumped up and clapped his hands. Lashed to the side of the boat was a slick, thick, black shape that ran its entire length and a bit more. The blackfish of Barrouallie. A charge of excitement flowed across the beach as the villagers drifted into the water to greet the whalers.

Esther Irene never once took her eyes off the whaling boat as it slid into the shallows. Even though she didn't seem to be doing anything, it was obvious she was there in an official capacity. The whalers cut the lashes, and the whale rolled heavily into the water. Esther Irene watched carefully. Other villagers, mostly the boys, took ropes and hauled the large carcass onto the beach. Esther Irene was right there beside it. The two-man crew unloaded the boat and broke down its gun. All of the gear was carried to a shed off the beach. Millington stood beside Esther Irene as she took a large knife and cut shallow, crosswise slices the length of the carcass. Then Millington took his cutlass and cut the slices through, turning the great black fish into red slabs of meat. Esther Irene directed different men to haul off particular sections of meat. The skull was carted away for its oil, and it occurred to me that this meat would be in the cook shops that evening or marketplace tomorrow.

Esther Irene's official capacity was obviously the broker. She pulled an oily clump of bills from her pocket and counted out a number of them into Millington's hand. By this time his crew had returned to take the boat back out to deeper water, where it would remain anchored for the night. The one I would come to know as Veron, the tall one, still carried the harpoon with him. Veron at seventy had leg muscles carved out by seven decades of demanding work. The harpoon he held had a shaft thick around as a three-year-old hickory sapling with a large barb off the hook that would make it impossible for the whale to dislodge once it was set. All the while nobody paid any attention to me whatsoever, not even the children; but, while Millington counted out bills for each of his crewmen, Esther Irene said something to him and nodded in my direction. He did not look at me, and neither did they, but the probability was that they had checked me out and sized me up even before they came ashore. Although no one had yet acknowledged my presence, a white man in a black village is not exactly anonymous, and the whalers, being hunters, would have noticed everything about me. When the crewmen pocketed their money, Millington finally turned and walked over to where I stood. I was only a few feet away across the sand. Both crewmen stayed where they were and watched. The harpooner rested with one foot up on his harpoon, like a Masai warrior leaning on his spear. All the people on the beach stared as Millington approached. I felt stripped and dissected by every eye in Barrouallie. Hester Prynne on the scaffold with the scarlet letter blazing on her chest could not have felt more scrutinized than I did.

I put out my hand. Millington took it. The veins in his forearm stood out in relief, and his hand was rough to the touch though his grip was surprisingly soft.

"I'm Millington. Miss Irene say you want me."

He didn't smile. He didn't try to ingratiate himself in any way. He spread his legs, folded his arms, rooted himself in the sand, and listened with his head down. His crewmen took it all in. I felt as if I had very little time to explain who I was and what I wanted, as if I had only seconds to be rejected or accepted by these men. I put my case briefly, to the point. Would they take me with them? Who was I Millington wanted to know? I had hunted all my life, I replied, and I was interested in men who hunted the way they did. I was also a writer, and I wanted to make a record of what they did because after them nobody would be doing it anymore. It was his boat said Millington. He didn't mind, but his crew had to approve, too. He called them over. The one with the harpoon, Veron, towered over me. The other, Aaron Moses, stared at me as if we were prizefighters listening to the referee's instructions.

Millington's accent was so thick and his words tumbled out so quickly that I could barely understand what he told them. There was not so much as a wrinkle of a smile on any of their faces.

"De watah deah, sometime de watah rough," said Aaron Moses.

"Yes," I said, "I expected that."

Veron shrugged.

"We'll take you," Millington said.

Veron ground the butt of his harpoon into the sand.

"Can I see it?" I asked.

He barely nodded and let it go. It fell toward me. I grabbed its shaft, hefted it, and was staggered by the weight. Veron's facial expression did not change. I figured it must have weighed at least fifty pounds and later learned it was sixty.

"*Fawr in de mawnin',*" Millington said. I said I'd be there.

At 4 a.m. the next morning the moon was still high and big over the harbor. She was full, a radiant disc of white light mirrored in the clement water of the Caribbean Sea. An hour earlier I had kissed my sleeping wife and son good-bye, packed a waterproof bag with camera, film, bright red life vest, two freshly baked rolls, and a can of juice, and driven the narrow coast road in the dark. There were no street lights on St. Vincent. Now, I stood on the jetty awaiting the arrival of Millington and his men. Small waves lapped gently against the pylons. The silence was profound, so deep I imagined you could actually hear someone's thoughts. Millington showed up first.

*"De watah calm,"* he said. *"Fish be comin'."*

I heard a splash, looked over, and saw Aaron Moses swimming out toward the boat. Veron lugged essentials from the storage shed and arranged them on the beach—an outboard motor that he carried over his shoulder, a can of gasoline, coils of rope, the harpoon. Such strength from a man of seventy years—such strength for anyone! The man was a match for no one. Aaron reached the boat, clambered aboard, put the oars in the locks, and rowed her to shore. He jumped out when her belly touched the sand, and he and Millington and I skidded the bow of the boat onto the beach. Veron returned with the mast over his shoulder. I can only guess at what the mast weighed, but I'll bet it was at least twice his weight—a solid shaft of wood that was a good eight inches in circumference and fifteen feet high. Veron waded into the water with it, and Aaron helped him lay it flat in the boat. Millington attached the outboard. Aaron unwrapped the chamber, butt, and trigger mechanism of the gun and put it together atop the iron quadropod on the bow. He swiveled it from side to side. It moved easily. Veron finished loading the remaining gear, and the four of us pushed the boat free of the beach. Then Millington motioned me aboard and bade me sit in the stern on the starboard side.

He stayed on the gunwale on the portside of the outboard, Aaron Moses stood on the bow behind the gun, and Veron sat on the mast housing plank in the middle. Neither of them said a word to me. I don't believe they had even looked at me. I'm not sure if Millington did, either. Maybe he simply pointed out where I was to sit. And stay. These three men were as fierce as any I'd ever been with, more so—men who could not be tamed, men savage and feral, men who took on the largest creature on earth. I admit to feeling unsteady, because of the boat, yes, a little, but more because of these men. I felt it: there was danger all over this adventure. I can't say I felt fear, but I didn't feel totally collected, and it wasn't only because of Leviathan.

Millington pulled the starter cord on the outboard. It roared to life. He guided the vessel out to the sea. It was five a.m. The water had swallowed up the moon. The sun was just beginning to glow, and the sky was the color of strawberry ice cream. This was the moment I had dreamed about for ten years. The hunt was underway.

The shore was a faint, gray line in the distance when Millington cut the motor. *Faith*, as the boat was named, rocked gently in the water. It was the only sound to be heard, but any feeling of peace I had was now replaced with an air of anticipation. Something was bound to happen, but what? And when? As the sun climbed in the sky and burned off the haze of morning, its glare upon the water intensified into sheets of bright light. It became impossible to keep my eyes open without sunglasses, yet Millington, Aaron, and Veron scanned the surface of the sea searching for sign without so much as a squint. They could tell, simply by the way the water moved, what swam beneath it.

"Shark," called Aaron Moses as he traced a path with his finger just in front of the bow.

I did everything but jump out of the boat, and still I couldn't see a thing. One flying fish, then two, then three leapt out of the water on the port side and skimmed for yards before darting back below. Millington mumbled about something feeding on them and that's why they fly. None of the men said a word to one another. This crew had been together so long they didn't need to. Veron bent to lift the mast. The muscles in his calves and thighs bulged with the weight. Aaron turned from the harpoon gun to help him. They maneuvered the foot of the mast to the lip of the hole in the housing plank, walked it up, and dropped it in with a solid *thunk* of wood against wood. Veron unleashed the ragged sail. Instantly the wind filled it with a loud *thwop*. It billowed out, its patches holding, and *Faith* slipped lightly through the water. In the span of a few minutes, the shoreline had totally disappeared from view. There was water, three hundred and sixty degrees of it, and sky, and a horizon that never came any closer. I had entrusted myself to the seamanship of a man I had only met a little more than twelve hours ago—an unknown sailor and a leaky boat. The thought went through my mind that maybe this wasn't such a good idea after all; but, as soon as it did, Millington slapped me on the shoulder and pointed to a spot off the stern. A whale was sounding— a humpback—probably a quarter mile distant. I turned in time to see the great tail slide into the water, its flukes silhouetted against the sky. It was the first time I'd ever seen one. Even at a distance it seemed huge to me; and, yet, the sea had swallowed it up so easily.

"My God," I thought. "It's there, in this ocean, and so am I."

Veron reset the sail. Millington adjusted the rudder and fixed a course toward the spot where we had last seen the whale. Aaron stood on the bow, slipped off the gun's safety, and lined up the deadly point of the harpoon. In that split second, so many disparate feelings surged through me. I wanted to get closer to the whale, wanted to join in combat with such a colossal creature, but to do so probably meant its death and possibly my own. Millington and his men were here to slay the giant in order to ensure their own survival. They were not here to witness a primal encounter but to engage in one. Without death, the experience would be incomplete. Without death, there would be no triumph. There would be only the punishing hunger of unbearable poverty.

Aaron Moses sang out, *"Big one deah. Come up. Come out de watah deah."*

As we approached the spot, Veron trimmed the sail and allowed the current to carry the *Faith*. We waited. The sea absorbed all our concentration. I got my camera ready and shot pictures of the crew in the meantime. For all the attention any of them paid me, I might as well have been a barnacle on the hull. Somewhere in the millions of gallons of water beneath us swam the focus of all our attention. I remembered medieval woodcuts showing whale-like monsters savaging the vessels of helpless seamen and then plucking them from the water by their heads, and I thought about how easily the humpback could surface below us and capsize our modest boat. I considered taking out my life vest and putting it on, but nobody else wore one, so neither, I decided, would I. It might seem rather harebrained that I would risk my life rather than risk their contempt, but, at that moment, I had another point of view. What was to keep them from tossing me overboard? Who'd know? As for the life vest, my wife had insisted I buy it and implored me to wear it; but, at that moment, while my mind shuffled through all the perilous possibilities of what might happen, most of me believed that none of them really would. And if one did? I hoped that she'd forgive me, and that I'd be someplace where I could get the message out.

We waited patiently over a peaceful sea. Nothing breached the surface of the water. After all, this was not the thundering surf of the North Atlantic, nor the South Polar Sea where a man overboard would freeze and die in seconds. It was the Caribbean, warm and soothing to the touch. Hurricane season was still months away. The sun glittered on the water. The *Faith* rocked gently as a cradle and nearly lulled me to sleep. We waited. I was in a kind of daze when I realized something had changed. A wisp of cloud, powered by an easy wind that wasn't there a second before, passed in front of the sun, and it lost some of its shine. The air temperature seemed to drop, not much, but perceptible. Our bow rose and fell then rose and fell again and again as a series of low-level swells passed under the hull. On the way out here the *Faith* had taken in a few inches of water; but, until this moment, the crew had not bothered with it. Now it was sloshing around our feet, and Veron began bailing it out with an empty gourd. The wind came on a bit harder, rattling the hardware on the mast and snapping the edge of the sail with the sound of a wet towel flicked against a tile wall. A wave rolled in from the starboard side, my side, and pitched the boat over so far to port that the gunwale nearly touched the water.

Millington tried to start the outboard. It coughed and died three times before it finally caught. He brought the boat around as Veron stopped bailing so that he and Aaron could muscle the mast out of its housing and lay it flat. I stowed my camera and took up the bailing to be of some use. Aaron slipped and went down under the weight of the mast. His leg twisted under him, and he caught the full load on his shoulders. A little to the left would have split his head; but, as it was, he shook off the wallop like a full-back shakes off a tackler and helped Veron settle the mast securely against the bulkhead.

When the rain came it came on fast, like rain always does in the tropics. It doesn't sneak up on you, doesn't tease you with a light drizzle. It just unloads, all at once, like a fighter with a sneaky right hook. A freak storm assaulted us with a hammering downpour suddenly driven by a hostile wind. It came from every direction; and, since the *Faith* was an open boat, there was no way of avoiding it. We were totally exposed to whatever the weather had in store. Waves broke over the boat and bombarded us with spray. In seconds I was as soaked as if I had jumped into the sea itself. Bailing made no sense now because the breaking waves dumped more water in than we could possibly bail out. Veron and Aaron Moses hunkered down under sheets of torn and tattered plastic. Millington stayed in the stern manning the tiller and coaxing the outboard. He concentrated on the sea ahead of him and struggled to keep a steady course. He seemed to know where he was going though I don't know how because he had no compass and there was no land in sight. I had no raingear, so I hunched over on the other side of the outboard and made myself as small as possible. I will get through this, I told myself. This cannot last forever, and I will get through it. Just then a trough opened up in front of us, and the bow plunged down into it. Almost immediately, the stern reared up in the air as a roller heaved under. I grabbed the gunwale to keep from being tossed overboard. This was angry water. It churned up foam and spray. No green, no blue, no clear, just white, hissing, punishing water. Then the bottom dropped out from under us like a trapdoor on a gallows, and the *Faith* plummeted straight down, slamming with the full force of her against the hard floor of the sea. Walls of water poured down on both sides of us, but the *Faith*, thank God, was dauntless. She shook off the water and shimmied to the surface as the bottom filled in. A nanosecond later, when I thought it could get no worse, my guts lurched and I knew, oh, my God, how I knew! It hit me with no warning. I realized it before my brain could even form the thought. I had never felt it before or since, but there was no questioning what this was. I was seasick. My insides kicked into spasm. It was awful. All I could think of was, "My God, don't let me throw up in this boat!" They'd toss me for sure.

Again we plunged and heaved, and again a wave broke over the boat. I must have had my mouth open because this time I swallowed a bucket of salt water. I gagged but held it in. When would this be over? Soon, mercifully, we reached shore.

The next day, again at 4 a.m., Millington hitched a dragline off the stern once we were under sail. When he spotted the school of skipjack he sailed right into them and hooked one almost immediately. The fish was dark on top with a silver underside; but, aside from a single leap, it didn't put up much of a fight as Millington pulled it aboard and killed it with a wooden club to the head. Millington quickly unhooked the fish, rebaited the hook, and returned his line to the water. Aaron Moses was behind the harpoon gun swinging it from side to side as he tried to get a bead. The harpoon gun went off with an explosive crack, and a scarlet patch appeared in the water. The dead skipjack left a ribbon of blood in its wake as Aaron hauled it aboard. A skipjack (a species of tuna) is only about three feet long, so I was surprised that he even hit it given the primitive quality of his weapon. He pulled out the light harpoon, and Veron took over. Veron held both fish over the side of the boat and butchered them so the blood would flow into the water. He quickly cut them into chunks and tossed them back in. By this time, Millington had dragged a third one over the stern. Veron clobbered it and butchered that one, too. He left its meat bobbing beside the boat and its blood drifting with the current.

After butchering the skipjacks, Veron and Millington trimmed the sail so that we would drift along with their remains. Millington baited another dragline, this time with a small pogie taken from the belly of one of the skipjacks, and Aaron reloaded the harpoon gun. We were back to waiting. Aaron, on the bow, sang out:

*"Mongoose, mongoose, come out de watah deah. Come out de watah. Roll ovah."*

The big harpoon still lay unused where it was first stowed by Veron. Millington stared after the slack dragline. Blackfish can run in schools of hundreds or even thousands, but also pods of a dozen or less, and I imagined them, exceptionally social animals that they are, signaling one another by clicks and whistles, coursing like underwater wolves after their prey. I knew that fierce hammerheads and dog sharks, which travel and hunt in packs like wild dogs, were out there, too, as were bull sharks with heads wider than long, the most common attacker of human beings. Millington, who does not shy away from a lethally wounded, sixty-foot, 80,000-pound whale, professes not to be troubled by sharks, either, but he has dreamed of being ambushed by one as he puts bloody chum into the water. The sea is smooth and calm, and then the monster's jaws suddenly rupture the surface and rip off his right arm at the shoulder. In the dream he remembers his detached arm quivering in the shark's jaws as if the thing were waving good-bye. "Like a horra pitcha." He laughs when he tells me about it—horror picture, indeed.

The mongoose, too, is well suited for hunting. Millington told me it sometimes even swims in rows like soldiers, and he has seen it attack dolphins and the calves of sperm and humpback whales in the open sea. He has seen it eat the tongue right out of a living whale. Its sharp, conical teeth are capable of tearing apart all kinds of fish, and its throat is large enough to swallow them. Millington has also witnessed them in a skirmish line, maybe eight or ten of them, actually zeroing in on a lone shark. This time it was the shark that was prey. As they closed in on the shark, they shifted formation, spread out, and encircled it. No escape. The shark was trapped. The attackers launched a lightning attack and closed in on their victim quickly.

It only took seconds to destroy it. When the mongoose were finished with it, they regrouped in formation, basically, a straight line, and swam away. They didn't eat the shark. They just killed it.

Mongoose is a formidable predator, but there is nothing on record or in the folklore of the trade about them attacking and killing humans. Millington does not dream about the mongoose.

My thoughts were full of sharks that day, so much so that I was completely unprepared for what happened next. The dragline off the stern went suddenly taut. Something had taken the bait. Millington immediately gave the line a hard jerk to set the hook. The line stayed taut. Whatever was there had gobbled the *pogie* and was caught. Veron and Aaron stopped what they were doing and watched as Millington began to haul the line in steadily hand over hand. The line slackened for a second then went taut again as the fish fought back and took it straight down. Millington quickly paid out the line, letting the fish have it all. The line went nearly vertical. It was so taut you could *twang* it. The fish was almost straight down below the boat when it shifted gears and swam away again. I moved to the center of the boat to give Millington more room.

"Dolphin," Veron said.

I didn't think I heard him properly.

"What did you say?" I asked.

"Dolphin," repeated Veron and indicated the line with a movement of his chin.

This just did not compute. It was common knowledge that dolphins were regularly caught by accident in gill nets set for tuna, but who actually fished for them?

Who deliberately baited a hook and rigged a line for dolphin? Obviously, the crew of *The Faith* did, although it might as well have been a mackerel or a cod down there, some commonplace fish that schooled in millions, for all the emotion they showed.

The dolphin had taken the line all the way out to the end without rising to the surface. I marveled that such a slender line should have the strength to hold a small whale fighting to break free, for that's what dolphins are, small, toothed whales, card-carrying members of the *cetacean* family. Then Millington leaned back on the line and resumed hauling it hand over hand, and I marveled at the strength of the fisherman, too. There was no slack in the line, yet the dolphin was coming in. A dolphin may be small for a whale, but it still weighs hundreds of pounds, a thousand even, certainly a match for a man with a line; and, yet, the dolphin really was coming in . . . until it wasn't. This one had a strong will of its own. It streaked back out to sea, dove on a dime, and took the line down, then turned and rocketed back up toward the surface. I knew this, of course, by the play of the line that went straight out then down and tight then slackened again.

"Fish be comin'," said Veron.

It sure was. It broke the surface and launched itself into the sky. Drops of water popped off the line and caught the spectrum of the sun like glass beads. What was on the hook was not what I expected. It wasn't a dolphin at all, at least, not what I thought of as a dolphin. Instead it was the most exquisite fish I had ever seen, a creature conjured up by a wizard, a torpedo of peacock blue and silver yellow with a forked tail of flashing gold and translucent, copper red fins like wings. It arched above the sea then all three feet and more of this fish reentered the water with barely a splash only to immediately propel itself into the air once more. This time its blazing leap must have carried it forty feet. It simply radiated light, and I stood mesmerized by its uncommon beauty.

Millington played the fish until he could bring it close enough to the boat to crack it senseless with his club. I helped him haul it aboard, not because he needed any help from me but because I wanted to touch such a glory of creation. My guess is it weighed about forty pounds. In death, its colors began to fade immediately. Veron gutted and filleted it right away. He sliced off pieces of firm, pale pink flesh and handed them around. Even though I don't like to eat fish, especially raw fish, there was no choice but to taste this one. Veron was treating me as one of the crew, and I had better act like it. Without hesitation, I put a piece in my mouth and chewed slowly. It had a delicate, almost sweet flavor, really quite delicious, not at all fishy. Veron handed me another piece, and I ate that one, too. Later I learned that what these men called dolphin was really a fish called the dorado, a gourmet treat served in fine restaurants as mahi mahi. The species of mammal I thought of as dolphin they referred to as porpoise, regardless of how big or small it was.

Porpoises, as the fishermen knew them, came in the afternoon.

"*Pawpus*," said Millington, and an instant later at least a dozen dolphins, as I knew them, burst from the water on the starboard side of *Faith*. Another dozen or more appeared at port. I had seen dolphins before in other waters taking bow rides or following wakes, but that was only two or three, maybe four animals at a time. Now the sea teemed with them. There may have been a hundred. There may have been more. It was as if I were witness to a miracle, the world lavish with countless beings leaping from the sea. Music was playing, symphonic music, Tchaikovsky, the *1812 Overture*, and every time the cannon fired more dolphins would leap from the water. Every time the cymbals clashed another troupe would soar through the air. The drums would roll and the sea would open and with the exuberance of the truly free these living things would defy the gravitational pull of planet Earth. Port and starboard, bow and stern, they jumped and twisted and dove and jumped again. A full orchestra played on and on. There were scores of them, hundreds, wild creatures absolutely delighted with being alive. Was I dreaming? No, the dolphins were there, always just beyond my reach, but there, right there. It was a wonder.

So enthralled was I by the spectacle of dolphins that I forgot about the fishermen until the harpoon gun fired. I turned in time to see the last of the harpoon line snake out from its coil on the deck. One dolphin rose and spun partway out of the water then belly flopped down. Millington had maneuvered the *Faith* so that the dolphins had to cut across her bow, and Aaron had sent the harpoon home at a distance of about fifty yards. I couldn't tell where the harpoon hit, but I knew the wound must have been fatal because he was having no difficulty pulling in the line. Veron stood poised with his harpoon ready to strike if necessary, but the dolphin was already dead by the time it reached the boat. Millington, Veron, and Aaron Moses hauled nearly a thousand-pound carcass out of the water so that the tail was in the boat and the rest of the "pawpus" hung out limply over the water. Then Millington and Aaron held it in place while Veron sliced its belly open. Blood gushed out and formed a brilliant red crescent in the water as Veron quickly gutted the animal and returned its entrails to the sea. After this, all three men lashed the carcass onto the port side of the *Faith*. Once done, Aaron reloaded the harpoon gun, Millington returned to the tiller, and Veron bailed out the boat that had, as usual, taken water.

The crew had executed well. Everything had been accomplished efficiently and matter-of-factly. There was no joy and no sadness—at least, none that I could tell. Yet, it was hard for me to accept that this was only a job to them. After all, they didn't choose to be bait fishermen. They chose to go after the biggest game of them all. Didn't Aaron stand on the bow and sing to them? Didn't Millington work at hard labor for years while he dreamed of the day he could buy his own boat? Wasn't Veron, at seventy, still pleased with the way he could put the harpoon? These men were hunters; they carried themselves proudly. They were not like the other men on the island, the clerks, the van drivers, the crop tillers, the beggars. They didn't sell *ganja* or poach parrots from the forest. Maybe they felt joy was inappropriate when they took lives as big as the ones they had taken. Maybe, if they thought about it at all, sorrow was just too dangerous an emotion for a predator to have. Maybe a quiet satisfaction was as far as they went. People depended on them to kill well, and they had done that. Life was not a gift but a daily battle they fought with skill and determination. It was a very difficult thing these men did. Did they need to feel reverence, too? What more did I expect from them, and what right did I have to expect it?

Shortly after the dolphin was dressed out, the sharks came in. Veron picked up his harpoon and pointed out two gray-brown dorsal fins swimming alongside the carcass. Aron had the wooden club in his hand. He pointed off the bow to three more.

"*Hahmmahead*," Millington said.

Hammerhead. In an instant, it felt as if ice were packed around my heart. My reaction was completely involuntarily. I wanted to back away as I had trained myself to do when confronted with a deadly predator, but I was in an open boat. *Where was I going to go?* It was the first stage of panic. One hears stories of aggressive sharks attacking small boats. Not only was this a small boat, but there was nearly a ton of raw meat lashed to its bulkhead. I knew that the hammerhead is one of the most dangerous sharks to man. Some claim they're not, but you'll understand if I take Millington's word for it. His word? Actually, words: "a bad infliction." They are fearsome hunters who follow the blood trail with their peculiar heads swinging back and forth. My first clear sighting was of one lengthwise, parallel to the boat. It appeared as if it were some strange species of deformed life bred by a demented engineer. Most unnerving were the appendages that stood out from either side of its head, each with a bulging eye-ball at its tip, like the cue ball on a pool table. These things, these appendages were as thick around as the business end of Babe Ruth's bat. Of course, its head moved side to side as, I presumed, it searched for a target. I was both alarmed and enthralled. "Alarmed" doesn't quite do justice to what I felt. "Terror" doesn't either, nor does "fear" because the hunters showed none so I smothered mine—more like I was struck dumb. It wasn't the jaws that mesmerized me, it was those protrusions on either side of its head. They belonged to a creature to whom I was nothing but an easy kill.

The fishermen seemed concerned, but they didn't seem afraid. They'd been here before, and I hadn't. These were my first sharks in the wild. Veron stood and tracked their movements with the harpoon poised to strike. Stay still. Watch. Learn something, I told myself. Focus. Focus. This is not your fight. Be glad. Glad. Watching Veron calmed me down. I felt those sharks were in more trouble than I was.

The first strike came from portside at the stern. It was a sixth shark, one I hadn't seen before, another hammerhead. It erupted from below with open jaws, a cavern lined with sharp knives, and tried to rip the tail off the dolphin, but Millington struck down with a long, thick, automotive screwdriver and embedded the shaft deeply in the hammerhead's skull, dead center, between its two eyes. It twisted its head, wrenched away from him, and writhed and thrashed in the water with the screwdriver still firmly fixed in its skull. Veron got the second one, but not before its jaws locked onto the dolphin and began tearing off a hunk of meat the size of a beer keg. Veron thrust downward and pushed the heavy harpoon deeply into the shark's body at the place where its head connected to its neck. Still, the hammerhead would not let go. Veron leaned on the harpoon and pushed it even deeper into its body until it did let go, but its jaws continued to snap as it quivered on the harpoon. Blood spewed out into the water as Veron twisted and jerked the harpoon free, and then the shark sank slowly, even gracefully out of sight.

Nobody told the shark with the screwdriver in its head that it was dead. It continued to thrash about in agony as if it were having an explosive grand mal seizure a few feet away on the surface of the water. Violent convulsions shook its body. Its head flailed from side to side. It rolled over, revealing an off-white belly, then righted itself by lashing its tail. The shark was in its death throes, and its convulsions suddenly interested the others more than the body of the dolphin lashed to our boat. There were four of them out there not including the wounded one, and they attacked him as if they were a kamikaze unit. The hit was swift and pitiless. They tore chunks out of him, swallowed them, and tore out more. The water seethed and turned red until all that remained of the wounded shark was a blood slick. A second one must have sustained a bite of its own during the attack because the three that were left turned on him and shredded him, too. Shortly there was nothing left of either shark except for the slick of blood. One shark stayed at the surface snapping at the slick, but the other three disappeared. Soon he did, too.

*Faith* arrived back in Barrouallie at sunset. Esther Irene waited on the beach and supervised the butchering of what remained of the "pawpus" as she had done the pilot whale a couple of days before. I sent a small boy to the rum shop for four bottles of Hairoun beer, took one for myself, and handed the rest out to the crew. Esther Irene was annoyed at being left out, so I sent the boy for six more bottles. Millington's simple house backed onto the beach; and, once the gear was stowed away, he invited us to go there. We never went inside at all. Instead, he built a fire in a pit in the sand of his backyard, and we sat around drinking beer as the sun went down.

# SHARKMAN

### By Jerry Gibbs

*For thirty-five years, Jerry Gibbs was the fishing editor of* Outdoor Life *magazine. He retired and moved with wife Judy from northern Vermont to the coast of Maine. He writes as an introduction to this wonderfully weird story: "There is a story about the angler who went to Australia intent on fishing for great white sharks. Just before his charter boat sailed, the fisherman suddenly realized what had been bothering him. There appeared to be no bait at all aboard the vessel. The captain was quick to ease concern.*

*'Don't worry,' he told the sport, 'it's coming now.'*

*Indeed it was. Being led along the pier in hide that fit like oversize pajamas, plodded an ancient, swaybacked cow . . ."*

*In the tradition of Hunter S. Thompson, no story ever got weird enough for Jerry.*

They arrived in darkness. The air was heavy, thick with sea-wet. The two men climbed stiffly from their car after the long drive from the city. They walked with crunching footsteps in the gravel parking lot to the long boat. The boat was the *Huntress,* and she lay clean and white beneath night lights. It was forty-five minutes before they were to sail.

"You want some coffee?" Ed Hammond asked. "There's that diner we just passed. I could use a little walk."

"Sure," the other man said. "Just black."

"Be right back," Hammond said.

Tommy Wakeman watched his friend walk away. He had never seen him stay in one spot more than a few minutes except in a car or on the telephone at his ad agency. Wakeman envied his friend's mountain of energy.

Tommy stretched. His eyes felt gritty from lack of sleep, but otherwise he felt good. He looked forward to the day. Six years of selling magazine advertising space with its ritualistic business lunches was beginning to show around his middle section. Leaning on a strong enough fish or two would be a welcome workout. And kind of a nice way to wind down before the big change came. He thought a moment about Amanda, asleep in her small apartment back in town. They would be married in—he counted— just three days.

When Hammond returned, the two stood sipping coffee gratefully. Boat lights began to wink on slowly, one here, another down the line. There was pale light growing in the east and the stars glittered weakly, fading. The air was rich with pungent odor of marine organisms living and dead. They scanned the dock for some sign of the Captain. They needn't have worried.

Ten minutes before sailing time the parking lot exploded in the rattle of gravel hurled against wheel wells. Out-of-line headlights caught them as a battered pickup slammed through the lot and skidded to a stop. The engine died, then ran on dieseling before coming to ticking silence. The Captain unfolded from the truck cab looking waxen in the predawn light. Except for his height there would be nothing outstanding about the man to the casual observer. His straight, dark hair was streaked with gray. The hint of fast-growing beard shadowed his jaw. His eyes were what stopped you. They seemed overlarge for the man's angular face, so dark that iris and pupil flowed together, and they were flat, without depth. Perhaps it was just the light. Tommy gulped his remaining coffee.

Walking over, they greeted the Captain with enthusiasm. The man looked at them noncommittedly.

"You boys see my mate yet?" he said when the silence was about to become awkward.

"Don't think so," said Hammond. "He'd be down by the boat, I guess."

The Captain nodded. "Well, you'll see him." He was pulling things from the back of the truck. He looked up smiling with closed lips and Tommy thought he had never seen such a long slit of a mouth. His eyes fluttered. "You'll see him if he knows what's good for him," the Captain said.

He turned again, rummaging in the truck, turned and was in front of them suddenly with something swinging in one hand that made them take a quick step backward. In this left hand, held by the ears, were two dead cottontail rabbits.

The Captain headed directly for the boat, where a light had just flashed on. As though on cue, a man who must be the mate emerged from the cabin.

"What . . . what are those, uh, rabbits for," Tommy asked, keeping pace.

The Captain stepped on board. He half turned, mumbling something. It was more the sound of a feeding dog's growl when you've come too near the bowl. Then he ducked into the cabin. Wakeman and Hammond swapped wild glances.

"Morning," the mate greeted them. "How're you fellows this morning?" Compared to the Captain he was a shoulder hugger.

The cabin door opened and slammed. The Captain came out and climbed the ladder to the bridge.

"Name's Guido," the mate told them. "You fellows have everything?"

"Just got to get a couple rods," Ed said.

"Everything's here you'll want," said Guido.

"Well, I'll show you."

Hammond returned with two thirty-pound outfits rigged with leadhead jigs. "We'd like to troll out to where you start shark fishing," Ed said. "Maybe later even try sharks with these."

The mate was not enthused. "You might get away with it on the way out. Not sharking." He raised his eyes to the flying bridge. "We do it his way."

"Even if we're paying?" Tommy said.

"Yep. But we catch them when others don't."

Starter motors whirred, straining. The engine caught and rumbled alive. The Captain spread unzipped bridge canvas and leaned out. "Cast off," he told Guido. "We need to kill a shark."

Tommy sat shifting on a bare, side storage compartment watching the day brighten in watercolor blends of palest green, yellow, lilac that blushed into orange-red with the sun's approach.

Guido was cutting butterfish and menhaden into chunks and fillets for bait and chum. The mate worked methodically, slicing and chopping, a chef preparing some magnificent bouillabaisse. Guido kept some of the pieces in containers and scraped some into what was beginning to resemble thick soup. Some of the bait fish were left whole and placed into another container. When he was finished he tied a line to a pail handle, dipped up seawater, and began scrubbing down cutting board, deck, and transom. When everything was clean he dipped more water, swabbed the deck fresh, cleaned his brushes, and mopped and racked them. He arranged the chum cans and bait containers close at hand. Then he turned to the tackle.

There was a big 12/0 reel already rigged, but seemingly as a concession Guido went into the cabin and brought back four outfits with 9/0 reels and fifty-pound line. Looking up from rigging his lighter rod, Tommy shook his head.

"You'd think this was Cairns, Australia, or someplace," he said.

Guido smiled faintly, lifting his eyes to the bridge. "A little heavy for blue sharks," he said. "Maybe not for some other stuff that might be there." He began twisting a wire leader to the big snap at the end of the double line on the first fifty-pound outfit.

They headed for the five-mile-wide depression in the ocean floor that was known locally as "the hole." Less than two miles away, the boat slowed slightly and made a small course change. Taking advantage of the reduced speed, Hammond dropped his bucktail over and free spooled it beyond the wake. Tommy was only seconds behind. The leadhead jigs had silver Mylar in the skirts and green plastic teasers on the hooks. They hadn't moved a quarter mile when Tommy's rod bucked once, twice, then held bowed over while line vanished from the reel. He came back unnecessarily hard with the rod, cranked with no effect until the fish slowed. When it did, he lifted and moved the fish. He dropped the rod tip, clamping his thumb on the reel spool, and lifted again. Astern, a bluefish that would weigh well into the teens thrashed the surface but quickly sounded again. One the other side of the boat Ed was into another.

The boat continued doggedly on course; its momentum, coupled with the anglers' pumping efforts, soon brought the fish on top, slapping then sliding just below the surface toward the boat. Guido gaffed Tommy's fish, dropping it into the fish box, turning, hunting the other blue that had skidded to the starboard corner. The two fish drummed hollowly in the empty box while Guido swabbed the deck. The two men checked their terminal rigs, checked for nicks or frays, sent the now tail-torn lures back again.

"Nice fish," Tommy said. "Big blues."

"I don't mind starting the day this way," Ed answered. The impossible hour at which they had left was forgotten.

In less than a hundred yard's travel, the sun bulged over the horizon and both lures were hit hard again, almost simultaneously.

"There must be a big school of them out there," Ed yelled happily. The hooked fish cooperated, tearing off line in separate directions. Suddenly the pressure on both rods lessened though the fish were still there. The boat had slowed, then ceased forward progress altogether. In neutral, the engines rumbled cavernously, changing tone each time the boat wallowed and one exhaust broke the surface.

The Captain was before them. The scimitar-like knife in his hand looked like a slaughterhouse relic. He reached Ed's line first, pulling it down from where it left the tip guide, then slashing the mono with a quick upward thrust.

"What the hell . . . ," Ed started.

"We don't want those damned things," the Captain overpowered the protest. His voice was flat and dry. In a quick turn he was on Tommy's line, and the man's rod came back straight, a curl of monofilament hanging limply from its tip.

"What's the damned idea?" Tommy began, but again the Captain bulled in.

"You come to fish shark?" he demanded. The wind blew strands of his fine hair wildly like the telltale on a sailboat mast.

"Sure, but . . ."

"Then we fish shark." He pointed at their rods. "Put that toy stuff up. I'm almost there. We got a slick to start." He looked at Guido. "Check that kill stick and rack it behind the bulkhead."

The two men reeled their severed lines back on the spools. Ed shook his head. "The guy's nuts," he said. The boat turned southeast into the steady breeze and soon began to cross the lip of the hole. The rising sun that had shown so much promise slid into a featureless cloud bank, and now there was only intense glare and endless grayness to the east. By the time *Huntress* reached the offshore edge of the hole the wind had decreased. The Captain came around, idled checking their drift before he shut down the engines. There was still enough breeze so they would drift generally northwest, crossing the depression, climbing up the far bank.

The Captain leaned out from the controls. "All right," he said.

Guido scooped some of the concentrated mash he had prepared into the larger container that held some seawater and chopped menhaden. Then he stirred and began ladling the swill over. The ocean had gone oily smooth, matching the chum slick that began to coat the surface. There were only occasional heaving swells. With the slick going well, Guido started setting baits. Perhaps as a gesture for the lost bluefish, he set up only the fifty-pound outfits. Instead of the standard long-shank hooks there were giant tuna hooks on the leader ends.

The mate cut menhaden fillets, hooking two on a hook.

"A lot of them who fish makos like to hide the hook," he explained. "That's not nearly so important as giving a good size mouthful."

He stripped twenty feet of line from the first reel, wrapped a small plastic foam float to the line with a rubber band, then sent the baited hook back sixty feet from the boat. The second line was rigged the same way but set back at forty feet. The third line had no float. It drifted free just thirty feet from the boat.

Gulls screamed overhead, circling the growing slick. Soon the shearwaters moved in across the surface to skim larger bits of fish from the chum. They moved brazenly close to the boat. Guido yelled, waving his arms at them, but it did no good. Up in the bridge the Captain had removed the canvas doors and side curtains. He sat watching, a form without features in the shadows below the bridge overhead. No one spoke. In the silence, at wide-spaced intervals, a swell would come large enough to break past the boat with a hissing sound. And there were the cries of birds. Nothing else. Suddenly, the click on the port-side reel rattled with a short, staccato burst, jolting Wakeman and Hammond from torpor, onto their feet, hearts pounding. Guido was already at the rod, cursing. One of the shearwaters had hit the line, caught in it briefly. The Captain had not moved.

The day turned grayer still except for the sky directly overhead, which looked as though it were polished by a rarefied wind. Around them, though, the breeze remained gentle and the day grew hotter. The breeze squirreled occasionally into the east with little puffs that curled the stench of putrefying chum from the containers around in the cockpit. Guido kept the slick going. Occasional spatterings of the liquid hit the transom or gunwales, turning brown as they dried. The boat wallowed in the valleys of the swells drifting slowly, and dullness settled in, deadening senses. As they neared the inside slope of the hole, the bottom began its long, gentle climb to two hundred feet.

A quarter mile away and sixty feet beneath the ocean surface, one, then two beautifully sleek cobalt shapes turned to cut the drifting chum trail. From below, their undersides were the dead white of something that has never seen the sun. They were male blue sharks, and as they crossed the trail of chum again, their excitement grew, visible in quickening tail beats and body thrusts. A third shark joined them, and as a loose group they rose higher, gliding forward, hunting the source of the rich slick that signaled something helpless on which to feed.

More large pieces of fish flesh were suspended in the water the closer the sharks came to the boat, and the blues took them with quick snapping movements, rising, rising, until 150 feet from *Huntress* their fins cut the surface.

There was a thud of feet on the bridge deck. Tommy and Ed turned to see the Captain standing, his long-muscled arm pointing out straight. His mouth stretched in a wolfish grin. "They've come, boys," he said without looking down. "There they are." It was the closest they had seen him express something that resembled happiness.

Guido scattered two more ladlefuls of chum, slammed several butterfish overboard, then took two rod belt harnesses from their hooks.

"Unless we're into something really big we don't use the chair," he told the anglers. "Give you more fun with those blue sharks standing up."

From the bridge burst a rumble of deep laughter that bordered on the demonic. Ed and Tommy looked up but could see nothing. Guido continued to watch the sharks. Smiling his great slit-mouthed smile, the Captain went to a mahogany cabinet. He removed a bottle of blended rye whisky from its shelf, cracked open the screw top, and took two long, gurgling pulls. "So they came again," he muttered, as though it were a personal triumph. "They never get enough." He returned the bottle to the locker. A drop of amber distillate glistened in the beard stubble near one corner of his mouth. He wiped it with the back of a sun-pocked calloused hand, without brushing away the smile.

"They're here all right, boys," he boomed down. "Now get yourselves ready." Once again the deep laughter thundered at them.

Hammond rolled his eyes. Tommy just shook his head, watching the closing fins.

The blue sharks reached the middle bait, snapping small morsels of flesh.

"Lead fish is a damned big blue," Guido said. "He'll do nine feet, anyway."

As he spoke, the fish sounded and the cork on the middle line popped free. Immediately, the reel click went off like an old-time New Year's Eve noisemaker. As Guido lifted the rod to Hammond, who was closest to him, the fish came up again, moved to the float of the near line, and confidently ate it. Yelling, Guido yanked the line, which somehow popped free. "Get that line in and check if it's frayed," he yelled at Tommy.

Ed snapped off the click, jammed the rod butt into his fighting belt, and used his thumb to control the line that was running steadily out.

"Hit him now!" Guido ordered.

Ed slapped the lever over and came back hard twice. The rod bent and stayed that way. Line skidded from the reel, against the drag.

Another burst of laughter rippled from the bridge.

"Watch that far line now," Guido told Tommy. "You see him take that float? They do that sometimes. They take the farthest bait, then come in and eat the other baits or floats before you know it. They don't even know they've swallowed hooks," he said. "That line all right?"

"It looked fine," Tommy said.

"Now watch that far line," Guido repeated. "If his fish doesn't get in the way you may get tight."

The blue shark made two dogged runs out and down and Ed pumped it back nicely. Now it was closer to the boat, trying to beat straight down. The float on the outboard line began moving off. It had not popped but was skidding across the surface.

"There you go," Guido warned.

Tommy had the rod, and when he struck the float came free.

It was a smaller shark than Ed's but it ran strongly against the drag and Tommy let it go, enjoying the power of the run before trying to turn the fish. The first blue was at boatside now, spinning wildly. Guido grabbed the leader with gloved hands. Water exploded into his face.

"This is a good blue; you want him?" The mate yelled. The shark thrashed against the side. "We cut them off unless they're this big or you want the jaws."

"No, let him go," Ed said.

Guido ran his pliers down the wire, cutting as close as possible to the shark's head. The wire parted instantly and the shark rolled over once before gliding out of sight.

"Blues eat rotten anyway," Guido said. "We should get something else if you want to keep teeth. Maybe dusky.

Mako, if we're lucky. Don't count on whites. We'll keep any mako."

Tommy had worked his fish in now, and Guido took the leader.

"Cut it off," Tommy said. "I'll gamble for something bigger."

Again, the Captain was behind them, having appeared almost magically.

"First blood, boys," he said happily. "It's going to be a good day." He ducked into the cabin, reappeared with another rod and reel. "Another bait," he told them. "If they all go off together you'll have to dance."

Guido was already sending out fresh baits, keeping the chum going with his other hand. The Captain placed the bait on the new rod just slightly inside of Guido's sixty-foot float. There were four rigs set now, three with foam floats bobbing nicely. The sharks came again.

The first blue sharks, including the ones they released, had circled out, but there were others closer to the boat, sliding just below the surface, rolling easily. A grayer form with blunter snout appeared briefly near the float of the second line out.

"Dusky there," the Captain pointed. "We're on a roll, boys. Get them now before it dries up."

Before he could speak again, the float where the dusky had been disappeared. The Captain plucked the rod from its holder, swinging to Ed nearest him, thrusting the rig into his belt and hands with such force that Hammond felt as though he were impaled. When he set the hook the fish went down like lead. This was nothing like the fight of the blue shark. Ed spread his feet, tried to brace himself, tried to lean back but stumbled forward.

"Pull," boomed the Captain. "Break your backs, boys." He was not looking at them but watching the floats and the shark fins cutting the surface. "Pull your guts out boys, it's what you're paying for."

"More blues out there," Guido said. As though in response, the farthest float began to move off at the same moment the closest rig disappeared. Tommy saw Guido had the far rig and took up the near rod. Before he could do anything the Captain's voice hit him.

"Feed line," he ordered Tommy. "Don't put pressure on." He grabbed the rod he himself had rigged. "Damned blues, now," he rumbled. He pulled the float, trying to take the bait away from a blue shark near it, but was too late. The rod bucked. There were four fish on now, and bedlam reigned. The Captain responded like an over-wound spring, suddenly released. He screwed down the drag dangerously, came back on the blue so hard with a series of lightning fast jolts that the fish must have been pulled up and over like a chained maddened dog hitting the end of its tether. The small shark never had a chance. It was pulled in like some planning board, first on the surface, then just below. The Captain countered every sideways thrust it attempted, forced around its head, totally dominated the creature.

"I'm off," Guido yelled. His fish had broken free and the mate cranked his hook-less line in wildly. "Your fish still there?" he asked Tommy.

"I think . . . yes, there goes the line a little again."

Over in the port corner where the fish had brought him, Ed strained against the dusky, trying to bring the rod up to get line, watching what little he gained slip out again.

Teeth clamped, his face void of expression, the Captain bulled his shark in mercilessly, dropped his rod, held the leader with one hand, clipped the wire, then ran the rod and reel back out of the way near the cabin bulkhead. There was a bump then a thunking vibration from below decks. "They're on the props," he said.

Tommy looked at him wide-eyed.

"They'll try to eat anything in the water, now," the Captain said.

To prove him right, two blue sharks emerged from beneath the boat and ran its length. One of them rolled, looking up for a moment, its catlike eye just below the surface. Then it disappeared.

The Captain turned on Tommy. "Take that slack up, see where he is and hit that fish."

On the free line, Wakeman's shark had never gone from the chum slick. It was still on the opposite side of the boat from the dusky. Tommy took slack line in quickly, tightening up, then knocking the reel into gear. He hit the fish. It was maybe thirty feet from the boat.

In actual time, it took just seconds, but to Wakeman it was like a slow-motion film. The ocean surface blew apart in a perfect, foam-fringed circle, and something pale colored, something very big, rose into the air and seemed to hang there at the apex of its leap before turning over backwards, almost lazily, in a nearly complete flip. The creature crashed heavily back, slamming the surface like a concrete slab pushed off a three-story building.

The line burned from his reel with a dry, buzzing sound. He heard the Captain's voice crack one word. "Mako!" Tommy felt his line stop, sag briefly. He cranked furiously, then saw the clear nylon begin angling to the left. And then the big mako came up again. The fish boomed into the air in one head-over-tail leap, crashed back in, then exploded a second time into the sky. Again, it came up on a totally reversed course. It happened so quickly the huge boils from each leap were still visible on the ocean's surface before the shark crashed down the third time.

The mako veered to port where Hammond was still fast to the dusky, now farther from the boat. Ed felt the shadow of something over him, turned and saw the Captain. He saw the brightness of steel in the gray light, the quick movement as though a dark bird's wing had passed overhead. And then the pressure on his rod was gone. He looked up, pained.

"I want that mako," was all the Captain said.

Ed stumbled shakily toward the cabin, laid his rod against the bulkhead. He felt weak. Now it was Tommy's turn. The mako was on another tack, giving the angler a chance to gain line. It lasted only a moment. Again the fish ran, and Wakeman, on his feet, was dragged in a crab shuffle toward the opposite side of the boat. He came up against the unpadded wood gunwale, slamming his knees painfully, trying not to bend over.

"You lose that outfit and you're in after it," the Captain warned.

Long minutes passed in strained standoff.

"Stop resting him," the Captain said.

Tommy managed to raise the rod and move the fish. He dropped the tip, then repeated the action, mechanically. The mako came a little way then burned off again, taking all the gained line plus a little more. But there was no jump. Thirty minutes later Guido guided Wakeman to the chair. The relief to his legs was a gift. No longer was he forced to fight to keep erect against the mako's direction changes. The pale blue jeans he wore were blotted dark from blood around the knees where he hit the gunwales, but he felt no pain. What hurt were his arms, shoulders, and back. Especially his forearms. His mouth was drier than he ever remembered. Guido poured some water and held the cup for him to drink.

An hour later he was ready to do almost anything to end it and hide somewhere in darkness. Forty minutes more and the mako was coming in. It made small lunges now, then heaved halfway from the water in a final effort to leap. Tommy was beyond summoning a final surge of energy. He cranked and pumped raggedly like some antiquated machine. Close to the boat, the fish went down, but just a little way. Tommy raised him.

"Get up," ordered the Captain. "Move him along the side, and bring him in. Back yourself!"

Tommy obeyed, stumbling, not trusting his legs. Guido had the leader, held, lifted. The Captain reached with the flying gaff, sank the plow-pointed head deeply, and the shark went wild, spinning in a concussion of foam and water. The leader snapped. The line from the gaff head was tied to a bit on the boat, and when the fish quieted a little, the Captain pulled it close again. Guido used a straight one-piece gaff to bring the tail up. The Captain dropped a loop of heavy line over it and came tight. They held the shark shuddering there.

"It'll go close to four hundred," Guido said. The Captain went to the controls. He started the engines, threw over an unanchored marker buoy, then began moving ahead in a slow forward arc that took him away from the chum slick.

Hammond had taken the rod from his friend. Tommy sat dazed. Sweat matted his hair. His shirt clung to him. His face was slack. Hammond gave him water. When he was finished, he handed Tommy a beer.

"I can hardly hold it," Wakeman said.

"You did all right," Guido told him. He clapped Tommy's shoulder and the angler thought he'd fall over. Ed was talking when they both realized the boat had made a huge circle and was now headed back toward the marker. And the slick. From the primary controls the Captain turned. His eyes were bright and he was smiling, all teeth showing.

"We've got them coming boys," he said, his voice rich and full. "Now we have a mako. God knows what else will come."

No more, Tommy thought. I won't touch a rod. I can't handle anything more. Let Ed do it.

When they were back in the slick the Captain helped Guido lash the shark fore and aft alongside.

"We'll bring him in when something tries to eat him," said the Captain. "He's dead, but they don't stay that way much."

Guido started the chum again. Tommy looked over the side at the big mako, a nosegay of uneven teeth blossoming from its mouth, the flat dark eyes looking as though they had never been able to see.

Two more blue sharks came later. They made Tommy take a rod, and when he broke off, he silently thanked whatever fates were responsible. Later than afternoon, another dusky ate the twin fillet of menhaden and Ed worked him frantically, but this time there were no knives. When it became apparent that the fish was tiring the Captain sprang from his bridge aerie. He put Guido on the mako's tail rope. He took the forward line himself, and together they worked the carcass into the boat. It fell in, stiff and heavy, puddling gore, and Tommy moved quickly away. They raised the mako's tail section toward the superstructure to rest against the spar that served as a gin pole. The shark was then secured forward and aft. Then they turned to Hammond.

Along with fear of losing the fish, Ed sensed the lateness of the hour. He strained against the shark, anxiety whining in his head. But the dusky was coming. When it was over, the shark alongside, the Captain shouted before Ed could open his mouth.

"Gaff him," the Captain ordered Guido. "Don't turn him loose." He ducked into the cabin while the mate sank the clean stainless gaff hook. The shark's thrashing hammered the gaff handle violently, forcing Guido to fight for balance.

The Captain returned with the killstick. He dropped a charge in the powerhead, screwed it closed, and waved Hammond back. He went to the side, came over, two-handed, brought the 12-gauge head down surely. There was a thudding report. The shark shuddered violently, then was still. He had Guido raise the dusky halfway from the water, still keeping the shark over the side while he watched it.

Ed leaned against the far side. Finally satisfied, the Captain and Giudo brought the dusky over. It was a good fish, slightly lighter than Tommy's mako. The mate began hoisting the fish up, beside the mako.

"Wait," the Captain told him. He returned to the cabin, and when he reappeared he carried the two dead cottontail rabbits. They looked stiff and flat and totally out of place.

"Drop that snout," the Captain ordered. Guido lowered the shark's tail and went to its head. The Captain placed the rabbits almost tenderly in shadow on the deck. He grasped the wood handle in two hands and began working at the shark's mouth. The end of the handle burred and splintered. He dropped it in disgust, went back to the cabin rack, and returned with a harpoon shaft. With Guido's help he finally spread the jaws as far as possible. He took the first rabbit and thrust it into the dusky's maw. He took the burred wood handle, pounded the cottontail deeply down the shark's gullet.

Tommy and Ed watched, incredulously. The mate, who had learned to accept each new turn of the Captain's virtuosity, contained himself with effort. Now the second rabbit followed the first. The Captain plunged and pounded. Guido straddled the shark, straining at its snout. Gulls screamed. Lost in his world, the Captain worked feverishly, chortling, mumbling to himself. His hair flew like Maypole streamers. When it was done, he straightened.

"String him up," he told Guido.

The Captain's eyes slid like a lizard's over to Ed then back to Tommy. He mumbled something as he turned, then started laughing as he climbed to the bridge. It was not a totally unpleasant laugh. The two friends sat in disbelieving silence all the way back.

The precise moment that *Huntress* rounded into the harbor, the bank of grayness in the sky parted. The sun was orange-golden, sinking quickly, growing fatter, shimmering as it did. Above the dock, the evening crowd was waiting for the returning boats. There were children, mouths rimmed in residual ice-cream crust; the svelte, the bulbous, the old and the prime. As the boat turned prettily toward her slip, they pushed forward excitedly, murmuring. Some pointed.

The boat pulled neatly in, reversed engines, stopped dead. Guido was out quickly making lines fast. Then the unloading started.

The sharks were slid from the boat and winched high beneath the charter boat's sign across the entrance to the slip. The crowd grew louder as the huge fish inched to their final triumph. Tommy and Ed collected their gear. They paid the Captain.

"You'll want the mako," he informed Tommy. "Guido will steak him for you; ask him. The jaws are yours, unless you don't want them."

The fishermen agreed. They didn't know what else to do.

Then it began. The Captain strode from his boat. The late sun glinted from the large knife in his hand. He ignored the spectators, going directly to the sharks. He pulled up on the snout of the mako, revealing the awesome rows of violent-looking teeth. The crowd murmured again; fearful sounds. Children were lifted to shoulders to stare with eyes like small raisins in faces of whole-grain pudding. Some of them turned away.

The Captain went to work carefully on the mako's jaws. It was a bloody job, but soon he had freed the hinged cartilage. He set the jaws aside in a large pail, straightened, and for the first time acknowledged the spectators while wiping the knife.

"How'd you like to pay the dentist bill for one of these beauties," he said. There was a real titter of laughter. "Let's see what these fellows have been eating," he added.

Guido began trimming flesh from the mako's jaws. The Captain turned his knife to the thick dusky. The skin was tough but he worked through it, spreading and emptying the shark. A few of the spectators left. More refused to look—momentarily. Most of them pressed forward, and now one, then another pointed. Voices grew louder. There were a few unbelieving gasps. A child began to cry. The stiff forms of the wretched cottontail rabbits had been revealed. The crowd grew noisier. For one woman it was too much.

She was short, blocky without being fat, and her hair resembled a used SOS pad. She clumped down to the dock in a nimbus of indignity. She jounced right up to the Captain, looked down at the rabbits, then thrust her chin at the man towering above her. She gathered herself, took a deep breath, and in a voice that had ordered countless bowls of chicken soup for a generation of children and grandchildren, she spoke.

"Can you explain," she demanded, "how two land animals happened to be inside this . . . this fish?"

The Captain had played it perfectly, and now he approached Shakespearean eloquence. He looked once at the crowd, then at the woman. He swelled, readied himself.

"Madam," he said, "this is one of the mysteries of the deep we'll never understand."

The woman's mouth dropped. The crowd stilled for a moment before returning to its babble. The Captain strode from the dock and up to the parking lot. Crossing the distance to the truck, he began humming little bits of something that sounded, if you listened closely, very much like a verse from the Battle Hymn of the Republic. In his truck he slammed the door, opened a window. Now he was singing the words in a gentle rumble to himself. He reached into a small beverage cooler, plucked out an icy beer, and started off, still singing. He sang aloud, inspired, "His truth is marching on!" He drained half the beer, belched once, contentedly, wrapped himself around the steering wheel, and rattled off into the gathering dusk.